The Tales Teeth Tell

The Tales Teeth Tell

Development, Evolution, Behavior

Tanya M. Smith

The MIT Press
Cambridge, Massachusetts
London, England

This book was set in Stone Serif by Westchester Publishing Services. Printed and bound in the United States of America.

Library of Congress Cataloging-in-Publication Data

Names: Smith, Tanya M., author.
Title: The tales teeth tell : development, evolution, behavior / Tanya M. Smith.
Description: Cambridge, MA : The MIT Press, [2018] | Includes bibliographical references and index.
Identifiers: LCCN 2018005095 | ISBN 9780262038713 (hardcover : alk. paper)
Subjects: LCSH: Dental anthropology. | Dental calculus--Study and teaching. | Teeth--History.
Classification: LCC GN209 .S55 2018 | DDC 599.9/43--dc23 LC record available at https://lccn.loc.gov/2018005095

10 9 8 7 6 5 4 3 2 1

To my friend and mentor Donald Reid, for his steadfast support and inexhaustible love of teeth.

Contents

Introduction: Why Teeth?

Folks are often a little skeptical or politely suspicious when I tell them that I am a biological anthropologist who studies teeth. Perhaps fearing a quiz on their oral care, some promptly steer the conversation into seemingly more neutral territory, such as which classes I teach or whether humans are still evolving. I quickly parry that teeth contain detailed records of growth, health, and diet, as well as our evolutionary history. (And yes, we are still evolving—we'll get to that later.) Amazingly, *every day* of our childhood is "fossilized" during tooth formation, a record that begins before we are born and lasts for millions of years. That is, if we're lucky enough to not grind our teeth down during life. Humans wore through their choppers rapidly in the past, showing extreme wear in middle age or even by young adulthood.

So what are the tales teeth tell? The French paleontologist George Cuvier famously remarked, "Show me your teeth and I will tell you who you are."[1] As the forefather of comparative anatomy, Cuvier helped to establish how the shape of a structure relates to its function. The pointed canines of lions are useful for piercing animal prey, while the flat molar teeth found in specialized seed-eating monkeys efficiently crush and grind hard objects. Paleontologists use the anatomy and behavior of modern animals to infer the diets of extinct species, thus "telling" what something is from their teeth alone.

Cuvier also pointed to a deeper truth about teeth, albeit unintentionally, since the details of an individual's early life are registered in perpetuity. Teeth are unlike any other body part, recording physiological rhythms as often as every 8–12 hours, and as infrequently as every season or year. Kids today learn about annual tree rings in elementary school, yet many dentists and oral health specialists are surprisingly unaware that the focus of their livelihoods is a sophisticated time machine. Not to knock rings in

trees—especially given the important clues they hold about past climates—
but there is far more to discover inside our own mouths.

While this might sound like a plotline in the popular television show
Bones, my colleagues and I dive inside the teeth of ancient children to
establish how old they were. We employ painstaking approaches to coax
secrets from dental remains formed during the eras predating birth certificates. Counts and measurements of tiny time lines are more accurate than
any other aging method employed by forensic scientists. However, there is a
catch, as this technique only works up to a certain age. Once root growth is
complete—at around age 20 for modern human "wisdom teeth"—the addition of daily lines ceases, and these unique childhood records then gradually disappear with each meal or night of unconscious tooth grinding. A
small group of biological anthropologists mine this wealth of information to
understand how our ancestors grew up, how we evolved, and how cultural
transitions prior to recorded history have affected our health. In the following chapters, we'll explore the intimate precision, striking beauty, and
integrative power of growth rhythms in teeth (figure 1).

We will also consider the surprising records of behavior that remain on
tooth surfaces for millennia. For example, the plaque our hygienists carefully remove traps food particles, bacteria, and DNA from our own cells in
a sticky layer that can fossilize over time into dental calculus. While calculus doesn't show the same faithful time records as enamel and dentine,[2]
it captures human activity after our teeth finish growing, continuing the
story of our behavior and health into adulthood and old age. We'll learn
how clues such as microscopic scratches and pits formed during chewing
have spawned serious debates about the evolution of the human diet. And
we'll see how evidence from teeth may point to the uniqueness of our own
species, *Homo sapiens*, with our long childhoods, remarkably diverse diets,
and complex behaviors.

The Science of the Human Past

In 1859, Charles Darwin forever transformed biology with his theory of
evolution by natural selection, detailed in *The Origin of Species*. Although a
handful of naturalists and geologists had already come to understand that
life on earth evolves, or changes through time, Darwin conceptualized an
elegant mechanism for evolution that has since been subject to exhaustive

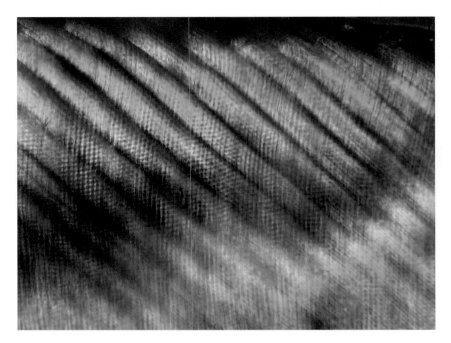

Figure 1
Growth lines inside a 9- to 12-million-year-old fossilized ape molar. Rhythms ranging
from 9 days (broad diagonal lines) to 24 hours (small, light and dark, box-like features
stacked vertically) reveal that this crown took more than 2 years to form. The
colorization is due to the use of polarized light microscopy, highlighting variation
in mineral structure. Fossil courtesy of the Natural History Museum (London).

scrutiny. He reasoned that evolution occurs through the greater reproductive success of individuals who are better suited for their environment than
others in a population. Later dubbed "survival of the fittest," this process
increases the proportion of organisms with anatomical, physiological, or
behavioral traits that help them survive and reproduce—provided that
these adaptations can be inherited. Darwin termed the mechanism *natural
selection*, which he contrasted with the artificial selection that occurs in
horticulture or animal husbandry though selective breeding. In later chapters, we'll explore how these cultural developments have left their own
impressions in the teeth of prehistoric humans.

Many biology students first understand how evolution works when
they're introduced to the Galápagos finches that Darwin encountered
during his world travels. Long-term study has shown that finches with

particular beaks have been able to colonize different islands and survive periods of environmental change. Beaks with sizes and shapes best suited to the local food supply are an adaptive trait that allows individuals who possess them to contribute their genes to the next generation. Birds with really mismatched beaks simply don't make it. Thus, populations of Galápagos finches evolve when individuals with genetically-determined adaptive features become more numerous than they were in preceding generations.

Darwin's friend and loyal defender Thomas Huxley extended this theory to explain human evolution in the controversial 1863 classic *Evidence as to Man's Place in Nature*, which Darwin followed in 1871 with *The Descent of Man, and Selection in Relation to Sex*. These books effectively launched *paleoanthropology*, a branch of contemporary biological anthropology that investigates human origins and evolution. Incredibly, Huxley and Darwin developed their ideas without much direct evidence of human ancestry. The fossilized remains of several *hominins*—humans and their extinct relatives— had been recovered earlier, but their discoverers, ignorant of the concept of evolution, regarded them as modern humans and relegated them to museum storage. The first fossils were unearthed from a Belgian cave during the winter of 1829–1830.[3] This discovery included a massive skull, a tiny upper jaw, and teeth from an infant Neanderthal, our brawnier, Northern-dwelling cousins (figure 2). We'll return to the story of this infant later, since cutting-edge imaging of its teeth has revealed exactly how old it was when it died. (Spoiler alert: it's younger than anyone expected!)

Paleoanthropologists who carefully pore over the teeth of ancient hominins are lucky compared to those who study other parts of their skeletons. Thousands of hominin teeth have been discovered and deposited in the natural history museums of Europe, Asia, and Africa over the past century. In some instances, fossil teeth are all that remain of long-extinct forms.[4] This is due to their highly resilient nature—more than 95% of the enamel cap is composed of mineral. It's not too much of a stretch to regard them as fossils in our mouths, although anyone who's had a root canal knows how alive our teeth are! Ongoing paleontological fieldwork uncovers new fossils each year, and this sometimes includes hominin children with pristine jaws. In 2006, my friend Zeray Alemseged vaulted to "rock star" status with his announcement of the discovery of a 3-million-year-old baby from Ethiopia (figure 3).[5] He had spent years carefully freeing the remarkable skeleton from a hard sandstone block, which the press dubbed "Lucy's

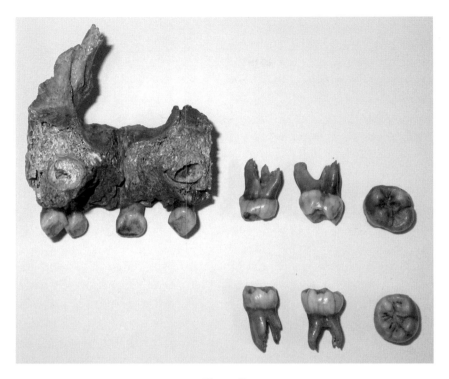

Figure 2
Infant Neanderthal upper jaw and associated baby and permanent teeth.
Individuals from this Belgian site (commonly known as Schmerling Cave
or the second Engis cave) were the first fossil hominins ever discovered.
Fossil courtesy of the University of Liège (Belgium).

Child"—a reference to the iconic female skeleton popularly known as Lucy
from the same species, *Australopithecus afarensis*.

Scientists today have an impressive toolkit available for the study of
rare fossils, and Zeray turned to my collaborator Paul Tafforeau for help
peering into this child's history. Paul is an expert in using high-powered
X-rays at the European Synchrotron Radiation Facility in Grenoble, France.
We have been working together for more than a decade to image the tiny
structures in teeth without needing to slice fossils open. Armed with many
terabytes of X-ray data and cutting-edge engineering software, our team
spent months working to determine how old Lucy's Child was when it died.
We'll return to the story of how state-of-the-art synchrotron imaging has
revolutionized the study of hominin development in the following chapter.

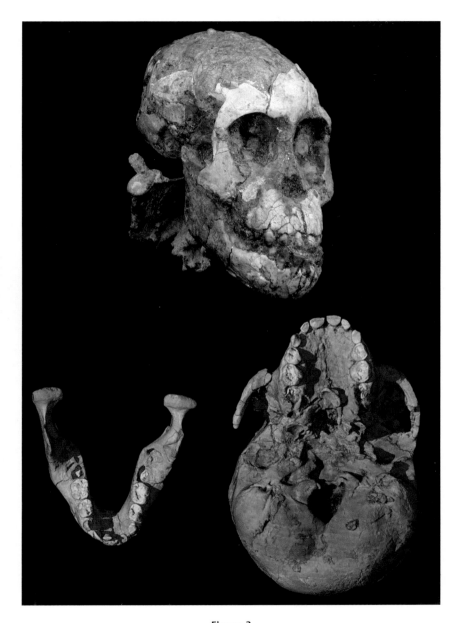

Figure 3
Infant skull from the most complete australopithecine child discovered.
Upper image is the original fossil; lower image is a virtual model showing
its erupted baby teeth after sandstone removal. Fossil courtesy
of the National Museum of Ethiopia (Addis Ababa).
Image credit: Zeray Alemseged and Paul Tafforeau.

What Makes Humans Human

In 1925, the young anatomist Raymond Dart announced a new ancient hominin, *Australopithecus africanus*, which means "southern ape of Africa."[6] Miners had recovered a child's skull from limestone cliffs at Taung, South Africa (figure 4), and Dart's interpretation of the fossil launched an intense scholarly debate. When scientists formally describe fossils, they employ Cuvier's comparative approach to identify similarities and differences with other closely related species. Dart compared the Taung Child to chimpanzees, gorillas, and humans, concluding that it was in an intermediate evolutionary position between living apes and humans. This was met with great suspicion by the leading anatomists and anthropologists of the time, who regarded the skull as much more ape-like than human-like—dismissing its position as a human ancestor. While other hominins had been recovered prior to the Taung Child, this was the first fossil that was different enough from living

Figure 4
The *Australopithecus africanus* child studied by Raymond Dart. The upper and lower jaws have a full set of baby teeth along with the permanent first molars. Fossil courtesy of the University of the Witwatersrand (Johannesburg).

humans to justify placing it in its own genus, *Australopithecus*. Given this unique position, it is little wonder that scientists initially struggled to understand its evolutionary relationship to living humans and great apes. In time, after considerable polarization of the scientific community, Dart's assertion was accepted. Numerous australopithecine species have since been found in fossil deposits. The acceptance of Dart's discovery also validated a radical proposition by Charles Darwin that the ancestors of living humans would be found in Africa. I'll discuss this further in chapter 5, where we'll encounter fossils that are even more ancient than *Australopithecus*.

Dart's description of the Taung Child includes an important observation about its teeth: "The specimen is juvenile, for the first permanent molar tooth only has erupted in both jaws on both sides of the face; i.e., it corresponds anatomically with a human child of 6 years of age."[7] Dart is correct; first molars in modern humans tend to erupt by about 6 years of age. Importantly, he identified a developmental yardstick—eruption of the first molars—that can be compared with other juveniles. The tricky part is determining whether the Taung Child died at the same chronological age as an ape or human. Think of the old proverb that 1 year of a human's life is equal to 7 years of a dog's life; most mammals grow up rapidly, achieving reproductive maturity and other life stages at younger calendar ages than do humans. Our first molars erupt around 6 years of age, while they emerge in African apes around 3 years of age. If the Taung Child was 6 years old when it died with its first molars newly in place, it had a developmental pattern like a modern human. But if it was closer to 3 years old, it more closely resembles the great ape pattern of dental development. Knowing this fossil's precise age would help to determine how rapidly its skull and brain grew, as well as whether it was likely to have had a long childhood and late maturation. Scholars debated both sides of this developmental conundrum heartily, and as we'll see in the following chapters, it took 60 years before two up-and-coming biological anthropologists settled the issue using the tiny time lines in teeth.

The Taung skull was also recovered with an exceptionally preserved replica of its interior surface, which has fueled studies of the evolution of the human brain. Our brain is often described as a "triune structure," based on three key evolutionary transitions. An ancestral reptilian core controls our reflexive fight-or-flight instincts, our deep mammalian centers facilitate the emotional attachment and regulation essential to parenting and group living, and our outer primate neocortex gives us the cognitive acumen to master complex social behaviors. It is this last evolutionary milestone that drives biological anthropologists to ask why humans are the way they are and when we became this way. Today, modern science has capitulated to the originally blasphemous ideas of Darwin and Huxley, embracing evolutionary explanations for our anatomy and behavior. Ideas about human evolution have also captured the imaginations of barefoot runners and inspired the popular Paleo Diet. The nascent field of evolutionary medicine has helped to explain health problems such as obesity and diabetes in the context of our evolutionary history. This is encouraging to those of us who've been confounded by deep resistance to the idea that humans descended from simian ancestors.

Scholars apply a similar evolutionary lens to understand ailments that befall our teeth. As fossil members of our genus, *Homo*, adopted tools and behaviors to process food, eating became easier. Because our muscles and bones respond to stress and strain, a reduction in chewing forces led to the development of smaller jaw muscles, and their bony supports followed suit by shrinking across generations. During the transition from hunting and gathering food to agricultural production 10,000–15,000 years ago, lower jaws became even shorter and broader.[8] This decrease in the size of our faces and jaws is frequently offered as an explanation for high rates of third molar impaction, driving the need to surgically remove them. Studies highlighted in chapter 3 reveal that molar impaction became more common over the past 10,000 years in small-jawed individuals. This potentially dangerous condition affects nearly one in four people living today.

Another lens through which we view human uniqueness relates to our behavior. As our recent ancestors perfected ways to cut, tenderize, and cook their food, they no longer required such perfectly attuned choppers to meet their daily caloric needs. This is particularly evident from skyrocketing rates of malocclusion, or misaligned contacts between chewing surfaces, in modern populations that rely on heavily processed foods. Teeth have also been probed for insights ranging from our prehistoric social groupings and stratification to our predilection for symbolic communication. In the final section of this book, I'll trace the origins of dentistry and highlight the handiwork of the first dental artisans. It turns out that modern humans aren't alone in their urge to floss or pick their teeth. Nor is a fascination with tooth necklaces unique to us, and in the past nothing barred some grisly ways of acquiring "dental bling." One type of behavior that does distinguish us from living and fossil primates is the worldwide human phenomenon of altering our own teeth—coloring, filing, and even removing them to fit in, or perhaps to stand out!

I taught at Harvard University for a number of years, and my favorite class was an annual research seminar that enrolled a small group of dedicated students. We'd spend three months using evidence from teeth to investigate the development, evolution, and behavior of humans and their primate relatives. Perhaps more importantly, my students were empowered to follow their own curiosity by performing research—a first for many of them. They would start with an exploration of living primates from the Harvard Museum of Comparative Zoology's impressive collections, examining oddities featured in chapter 4, like the "tooth combs" used for grooming and gouging bark. From this vantage point I would explain how tooth size tends to go hand in hand with diet and body size. Primates preferring a diet of insects, vertebrates, and tree sap tend to be small, while those above a certain weight rely on fruits, leaves, and seeds. The reason for the difference is a matter of metabolism and digestion. Small primates require high-energy diets that can be processed by short digestive tracts, while large primates bulk up on lower-energy items that they process more slowly in elongated digestive systems.

Humans, with our incredibly large, slow-growing, and metabolically demanding brains, have taken yet another tack. Hungry brains and an active lifestyle require a high-energy diet, and we can ingest and digest a dizzying array of foods. Our impressive set of teeth—an oral Swiss Army knife—includes a suite of tools adapted from our mammalian ancestors. Incisors help us bite into fruit, pointed canines and their moderately crested posterior neighbors pierce and tear meat and plants, and low-crowned molars help us grind up hard foods, such as nuts and seeds. The top selling point for our dentition is that it represents a balance between efficiency and overspecialization. Human incisors aren't as broad and shovel-like as primates that specialize in fruit; our premolars teeth aren't as crested as those that seek out leaves, insects, or animal prey; and our molars are not as flat and basin-like as primates that mainly rely on seeds. This dental blueprint suggests that, in keeping with our status as omnivores, we've hedged our bets evolutionarily.

In the following pages, we'll explore the human odyssey from a unique angle. Beginning with our development, continuing with our evolution, and culminating with our behavior, you'll learn how teeth illuminate our history like no other part of our anatomy. At times current and personal, these tales will engender a new appreciation of our remarkable evolutionary journey.

I   Development

I believe that much more of the intimate history of the individual is revealed to the microscopist by a study of the enamel than has been generally understood.

—Alfred Gysi, DDS, 1928 (as related by George Wood Clapp, DDS in "Metabolism in Adult Enamel," *Dental Digest* 37 (1931): 664–665)

1   Microscopes, Cells, and Biological Rhythms

Chewing over a few facts about dental development brings the fossil record to life in new ways. I learned this firsthand through studies of biological anthropology in college—as well as during my immersion in the field of microscopic imaging. Much of what we understand about teeth is due to the historic development of microscopes, which illuminate tiny tooth-forming cells and their mineralized secretions. Microscopic imaging has also taught us that teeth contain biological rhythms—internal clocks that mark each day of our childhood. This in turn has opened the door for anthropologists to calculate the age of ancient youngsters from their fossilized dental remains. In this chapter, I'll highlight several approaches to landscapes that are invisible to the naked eye—culminating in high-powered synchrotron X-ray imaging. This relatively new technique has allowed breakthroughs in fields ranging from art conservation to engineering. Learning how teeth grow will also elucidate numerous tales of our evolution and behavior in the pages that follow. For example, in chapter 6 we'll combine knowledge of dental development and X-ray imaging to unpack the origin of our exceptionally long childhood.

Geeking Out on Microscopes

Many of us met our first microscopes in an elementary or high school biology course, employing them for mundane explorations of onionskin or cells from a swab of our own cheek. Perhaps we were lucky enough to witness a fresh-water hydra regenerating, or the erratic movements of a protozoan zipping through a droplet of pond water on a glass slide. Like most teenagers, I took the objective lenses and eyepieces of the compound microscope for granted. I dutifully memorized its parts in order to pass my lab quiz without giving much thought to their significance. I couldn't have imagined the three-dimensional microscopic world that I would enter a decade later, nor the exhilaration of being the first person to explore the childhood growth of fossil primates long since departed.

During the seventeenth century, these simple tools supercharged the nascent field of biology.[1] The term microscope, meaning "small viewer," was introduced by a group of Italian scholars that included Galileo, who rose to fame through astronomical observations with his "far viewer," or telescope. A few decades later the British microscopist Robert Hooke proposed the word "cells" to describe the tiny units of life he encountered in the tissues of plants and animals. In his artful 1665 classic *Micrographia*, Hooke dramatically illustrated the jagged teeth of a snail as seen with a microscope, noting: "the animal to which these teeth belong, is a very anomalous creature."[2] Modern science has since confirmed that snails are unusual, as they continuously form new generations of teeth and modify the shapes of these replacements in response to changes in their diet.[3] Even more impressive are aquatic snails known as limpets, which grow tiny teeth made of the strongest biological material yet discovered. This design allows them to grind up rocks while feeding on particles that adhere to them. Engineers are studying these resilient teeth to learn how to create stronger artificial materials.

Hooke's Dutch contemporary Anthony Leeuwenhoeck may have been the first person to train his microscope on the structure of tooth enamel, reporting the presence of fine transparent pipes that look like small spheres when viewed end on. In the scholarly language of his time, he reports: "Six or seven hundred of these Pipes put together, I judge exceed not the thickness of one Hair of a Man's Beard."[4] Leeuwenhoeck's description of *enamel prisms* is quite accurate, and all the more impressive given the rudimentary microscope he employed. These long, rod-like structures are circular in cross-section, measure about 0.005 millimeters in diameter, and are nearly transparent due to their high degree of mineralization. I'll return to these important building blocks during our exploration of how teeth are built to last.

As someone deeply fascinated by the natural world throughout my childhood, I decided to pursue a bachelor's degree in biology at the State University of New York at Geneseo. My freshman-year advisor suggested that I enroll in a biological anthropology course, which I did, despite never having heard of the subject before. Exploring human and primate biology from an evolutionary perspective was immediately engrossing, and I realized that I had found an intellectual home. Like many young women taken with the biographies of Dian Fossey and Jane Goodall, I dreamt of going to Africa to study the behavior of mountain gorillas, and I diligently prepared to do so. Over the next few years I studied wild howler monkeys in Central America, tracked the secretive lemurs of Madagascar, and took numerous courses in biology and biological anthropology—an academic journey that led me back to the microscopic world I had prematurely dismissed.

Why would an adventurous woman give up her dream of studying gorillas in the wild to spend countless hours peering into microscopes in dimly lit labs? Like Robert Hooke and Anthony Leeuwenhoeck, I was quickly taken with a vivid world unseen by most. The romance began during my final semester of college, when I naively enrolled in an elective course on electron microscopy that happened to fit into my schedule. Led by the offbeat and enthusiastic Harold Hoops, a balding microscopist who actually wore a white lab coat to every class, I joined a small group of students learning to prepare, magnify, and photograph the internal structure of mouse cells. Hoops carefully walked us through the application of toxic chemicals that paradoxically preserve organ tissues for sectioning and staining. The process of creating our own glass knives and shearing ultrathin slices of rubbery blackened tissue intrigued me. We gently placed these slices in a transmission electron microscope, which was unquestionably the most sophisticated machine I'd ever been trusted to operate. As the beam of electrons passed through the tiny sliver of mouse spleen, something began to come to life inside of me.

The microscope monitor revealed a grainy black and white world that I'd only caught impersonal glimpses of from biology textbooks. Cell membranes, darkly stained nuclei, mitochondria, and other visual wonders populated the screen. I captured dozens of images of cellular landscapes, and then immersed myself in the art of creating prints in a darkroom. This introduction to microphotography sparked my deep appreciation of tiny biological structures. A faded black-and-white picture of a mouse cell that I magnified 78,900 times has hung beside my college degree for the past 20 years. As the first member of my family to finish college, I can't explain to them why this print means more to me than the diploma, nor why I've been chasing the perfect image through an objective lens ever since. The truth lies in that sweet territory where the boundaries of science and art blur, a private marriage between the left and right hemispheres of my brain.

Before I could don my black cap and gown to celebrate graduating with my college peers, Professor Hoops challenged us to design and execute an independent research project using electron microscopy. My mentor in the anthropology department, Bob Anemone, suggested that I investigate the microscopic structure of fossil teeth he had collected in Wyoming. It was no small task. I had no idea how to prepare the teeth for imaging, nor did I really know what I was looking for. At the time, I was vaguely aware that primate teeth had microscopic lines representing growth rhythms. Yet little was known about the structure or development of the teeth of other mammals, much less 60-million-year-old insect-eaters with teeth the size of pinheads. I passed many hours in the deserted basement of the biology building, searching for elusive growth lines at the helm of an electron microscope.

After days of scanning tooth surfaces at a maddening magnified crawl—hoping for some sign of their development—I brought an unusual abraded spot into focus. I immediately noticed long, thin, enamel prisms, and cross-cutting these prisms were parallel lines that resembled the daily growth lines of primate teeth (figure 1.1). I had done it! Accompanied by the drone of a vacuum pump and an electron gun, I had encountered a secret world in that windowless room. I captured a series of Polaroid micrographs from different angles and magnifications, and dashed across campus to show Bob. Capturing images of structures that had never been seen before was a tiny scientific discovery, yet it was the first electrifying experience of my academic career. A year later I enrolled in graduate school and began training in the rare art of coaxing primate teeth to reveal their microscopic secrets.

The Nuts and Bolts of Our Remarkable Dental Tissues

Teeth and bones make up the primary evidence for extinct vertebrates, including fossil hominins. This is because during fossilization they become even more mineralized, and are thus fairly resistant to decomposition and weathering. In order to explore the lives of ancient primates, I first needed to learn more about what teeth are made of and how they develop and function. The tooth crown, which includes the chewing surface, consists of a heavily mineralized, whitish coating of enamel on top of hard dentine and soft pulp (figure 1.2). Pulp is made up of cells, blood vessels, and nerves, and it's the site of occasional infections that are treated by the dreaded root canal. Dentine encloses this delicate tissue while giving roots their form, and an outer layer of cementum glues each root into tight-fitting sockets in

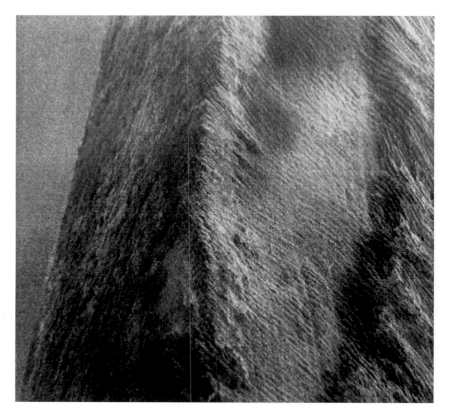

Figure 1.1
Enamel microstructure in a fossil tooth imaged with scanning electron microscopy. Faint vertical growth lines in the center of the image bisect diagonal enamel prisms. Fossil courtesy of Bob Anemone.

the jawbone. Enamel, dentine, and cementum are part of the skeletal system, which consists mainly of bone—the other highly mineralized "hard tissue" of the body.

Human tooth development begins before birth, during the first trimester of fetal growth.[5] Tiny clumps of cells lining the inside of the mouth, known as *epithelial cells*, migrate into the developing jaws—responding to signaling molecules that telegraph a great responsibility. These cells begin to stimulate deeper-dwelling *mesenchymal cells* to gather nearby, forming a dense cellular bud that marks the place where an engineering feat is soon to happen. As the cells divide and multiply, the bud enlarges, ultimately expanding into a bell-shaped structure that resembles the outline of the future

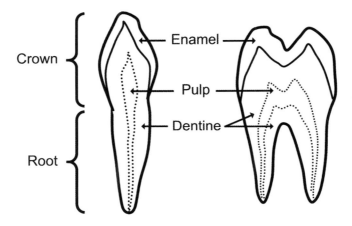

Figure 1.2
Tooth anatomy in cross-section. Illustrated with a single-rooted
canine (left) and a multi-rooted molar (right). Cementum is
shown in figure 1.6, below.

tooth crown (figure 1.3). A sketch rendered in the media of flimsy, transient cells foretells the genesis of something nearly indestructible.

I've marveled at how our body knows to form incisors and canines in the front of the jaws, and premolars and molars in the back of the mouth.[6] A series of creative experiments by steady-handed biologists deciphered this process more than 40 years ago. Scientists began to dissect and recombine living epithelial and mesenchymal cells from tiny mouse embryos, closely watching to see what kind of structure was formed. When early stage dental epithelial cells were paired with mesenchyme from other parts of the body, a tooth began forming. This wasn't so when epithelial cells from other parts of the body, such as the skin or limbs, were combined with early stage dental mesenchyme. No dental epithelium, no tooth. Yet the ability of the dental epithelium to initiate tooth formation is ephemeral—in mice it loses this potential after a few days, after which point the dental mesenchyme takes the lead in this cellular two-step. Moreover, these cells dictate the shape of a tooth, since transplanted molar mesenchyme creates molar teeth under incisor epithelium, while incisor mesenchyme will build incisors under molar epithelial cells. This works because genes in the mesenchyme turn on signaling molecules at specific times, telling the epithelial cells which shape to form.

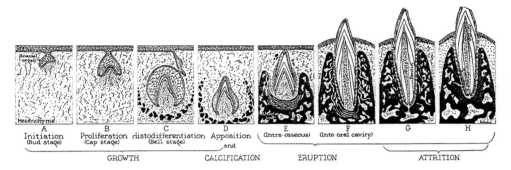

Figure 1.3

Classic illustration of the stages of tooth development and use. Progressive stages
from left to right: cellular organization (labeled growth), enamel and dentine
secretion (calcification), root formation (eruption), and wear (attrition). Black
regions represent the mineralizing jawbone, which develops around the tooth as
it is forming. Reproduced with permission from Isaac Schour and Maury Massler,
"Studies in Tooth Development: The Growth Pattern of Human Teeth, Part 1,"
Journal of the American Dental Association 27 (1940): 1778–1793.

Once the tooth type is decided, tiny signaling centers in the epithelium
called enamel knots get to work—ensuring that the number and placement
of the future cusps is correct before irreversible hard tissue secretion begins.
As the epithelial and mesenchymal cells take their places in opposing layers,
new signals direct them to restructure themselves into the main workhorses
of tooth development: enamel- and dentine-forming cells. Over the next few
months and years, these cells will undertake one of the most exacting tasks
in the body, continuously secreting their respective tissues while in motion.
This continues until the full three-dimensional crown is completed, leaving
enamel-forming cells on the surface and dentine-forming cells trapped in
the pulp space. Et voilà, the recipe for a tooth crown! It's a one-shot deal; there
is no possibility of a "do-over" if any hiccup occurs along the way. Because
teeth do not remodel once they are formed, developmental abnormalities
and major nutritional deficiencies are permanently recorded.[7] In chapter 3
we will learn about the most common glitches, which provide important
clues about stress, malnutrition, and illness during childhood.

My colleagues and I often analogize enamel secretion to walking backward while squeezing a tube of toothpaste, leaving a long thin trail that
reflects the speed and direction of the walk. But enamel-forming cells aren't

a closed system like a tube of toothpaste—they're constantly absorbing water and nutrients while assembling proteins. These proteins are secreted out of one end of the cell, where they help to bundle long, thin, crystallites into individual enamel prisms (figure 1.4). Such mineralized growth tracks are a permanent record of cellular activity—capturing enamel production with striking fidelity. In addition to being powerful cellular historians, these enamel prisms are remarkable for their elegant structure. They can reach over 2.5 millimeters in length—500 times longer than their width (figure 1.5). Prisms are also capable of withstanding high bite force and stopping the growth of microscopic cracks.[8] Their interweaving wave-like geometry can be compared to plywood, a design with alternating layers that prevent the wood from splitting when a crack is started by a sharp object like a nail.

After the crown is completely shaped, the enamel-forming cells stop secreting enamel, switching gears to pump out excess water and proteins while pumping in additional mineral components. This is the final step in calcifying the enamel. Mature enamel is the hardest tissue in the body, weighing in at more than 95% mineral, with the final few percent made up of protein and water. In contrast, the underlying dentine and surrounding bone are 60% to 70% mineralized when complete, retaining additional proteins and fluid that impart a degree of structural flexibility the enamel lacks. Scientists mine various elements and proteins to uncover childhood diets and ancient human migrations—more on this in chapters 7 and 8. In the final section of this chapter, you'll see how differences in mineralization impact X-ray absorption, which is important for radiographic and computed tomographic (CT) studies.

Dentine-forming cells also secrete a mixture of water, proteins, and mineral components. The cells deposit their rhythmic secretions inward toward the pulp and downward to form the root, ultimately encasing themselves inside the pulp space. Long, thin, tail-like projections of the cells mark their path, residing in hollow channels that perforate the solid mass of secreted dentine. Unlike enamel-forming cells, which are rubbed off the tooth surface once crowns emerge into the mouth, dentine-forming cells maintain their vitality. At a rate that is nearly imperceptible over weeks or months, these cells continue to secrete dentine along the inner walls of the pulp space, as well as in the channels occupied by their cellular tails. Thus older individuals have smaller pulp spaces and thinner channels—two telltale signs of age discussed in the following chapter.

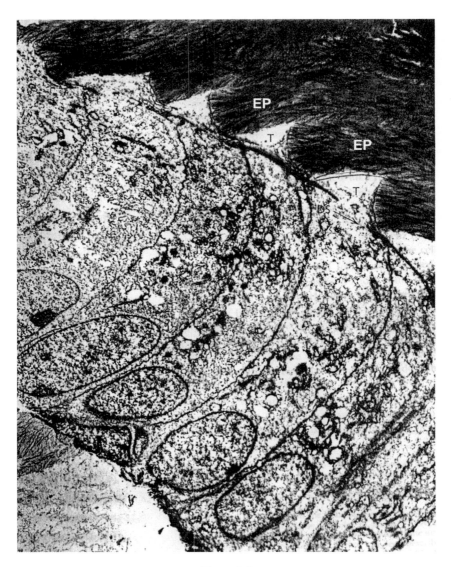

Figure 1.4
Transmission electron micrograph of enamel-forming cells secreting enamel prisms.
Enamel-forming cells run diagonally and secrete enamel prisms (EP) near the top
of the image, which are made up of thin crystallites. Darkly stained round nuclei
are visible on the opposite end of the cells. Modified from Erik Rönnholm,
"An Electron Microscopic Study of the Amelogenesis in Human Teeth: I. The Fine
Structure of the Ameloblasts," *Journal of Ultrastructure Research* 6 (1962): 229–248,
with permission from Elsevier.

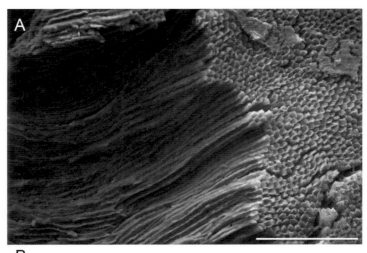

Figure 1.5
Magnified enamel showing the geometry of enamel prisms.
(A) Scanning electron microscope image of chimpanzee enamel showing the
three-dimensional nature of enamel prisms running from left to right. (B) Light
micrograph of enamel prisms (slightly curved vertical lines) running from dentine
(dark conical tip) to the enamel surface (upper boundary), a distance of more
than 2.5 millimeters in this human molar. Reprinted from Paul Tafforeau, John
P. Zermeno, and Tanya M. Smith, "Tracking Cellular-Level Enamel Growth and
Structure in 4D with Synchrotron Imaging," *Journal of Human Evolution* 62, no. 3
(2012): 424–428, with permission from Elsevier. Chimpanzee specimen courtesy of
Lawrence Martin.

The mature tissues that surround tooth roots are known as the
periodontium—giving rise to the term periodontal disease—which includes
cementum, gum, bone, and ligament. Our third and final dental hard tissue,
cementum, is a special biological glue. At first cementum-forming cells line
the growing tooth root, depositing protein fibers into the mineralizing dentine. This creates a tight bond between the dentine and subsequent deposits
on the root surface. The cementum-forming cells then secrete thin layers of
protein and mineral as they move away from the dentine. Eventually, fibers
of the periodontal ligament insert themselves into the mineralizing cementum opposite the root surface, creating a secondary anchor that sandwiches
cementum between the root and the ligament (figure 1.6). Cementum
is thought to be continuously growing in some areas, and it can thicken
later in life when teeth need reinforcement after forceful repetitive use. For
example, in chapter 8 we'll discuss how the extreme thickness of cementum
on adult Neanderthals' front teeth points to their use in demanding work.

Nature's Clock: Biological Rhythms

It shouldn't surprise me that just about everyone I talk to about my research
is aware that trees have rings that reflect time, but few people know that
their own teeth do as well.[9] Or that dental hard tissues are only one
part of nature's impressive system of timekeeping. Numerous animals and
plants form structures with regular time stamps, including bones, shells,
scales, insect skeletons, starch grains, and cotton fibers. In case you've ever
wondered why there are annual rings in trees, these are produced by seasonal activity in the cambium, a layer of cells that expands in response to

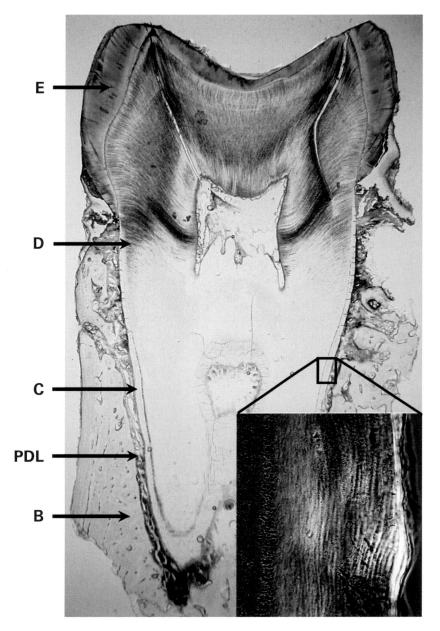

Figure 1.6
Molar thin section showing the hard tissues: enamel (E), dentine (D), and
cementum (C). Also included are the periodontal ligament (PDL) and surrounding
bone (B). Lower-right inset shows vertical light and dark cementum
annulations magnified with polarized light microscopy. Chimpanzee
specimen courtesy of Donald Reid.

changing day length, temperature, and rainfall. The natural world is full of rhythmic cues, including the phases of the moon and the rise and fall of tides. As life on earth evolved, it created its own internal rhythms. Tiny molecular clocks tick away faithfully in most living things. Interactions among genes, signaling molecules, and other proteins keep cells working on schedule and in concert with one another and the environment. Over the past century we've learned that both teeth and bones are formed by cells that squeeze out layers like clockwork.[10]

In order to read the temporal map inside teeth, scientists had to first determine how much time each type of line or layer represents. A pioneering Japanese team undertook this groundbreaking research on hard tissue rhythms during the late 1930s and early 1940s.[11] Masahiro Okada and Tasuku Mimura developed a powerful approach for studying growth lines in mammals. Laboratory animals were injected with small amounts of lead acetate and sodium fluoride several times over a number of days. They then sacrificed the animals and prepared thin sections of their teeth to view under the microscope. The injected biomarkers showed up as dark lines, serving as an independent external mark. The team compared the days between injections to the number of lines that grew over the same interval, finding a one-to-one correspondence that proved teeth have daily growth rhythms.

Despite the careful work of Okada and Mimura, there has been considerable debate about how to interpret these microscopic structures, and some doubt about whether certain features are actually formed every 24 hours. While I was working on my PhD at Stony Brook University, biological anthropologist Joyce Sirianni loaned me numerous monkey teeth that had been marked with fluorescent labels. She had completed an experimental study of their bone growth several decades earlier. Luckily for me, the labels also appeared in their teeth, which allowed me to validate the timing of different types of enamel growth lines. My dissertation committee felt that this was an essential step to take before I examined any fossil material. Without additional experimental proof, it was likely that some scientists would continue to question theories based on growth lines in teeth. After countless hours hunched over a microscope in a dark room, I again found myself experiencing an indescribable sense of elation as growth lines came into focus between labels. I counted 8 paired light and dark bands between two labels administered 8 days apart, confirming their daily nature (figure 1.7). The publication

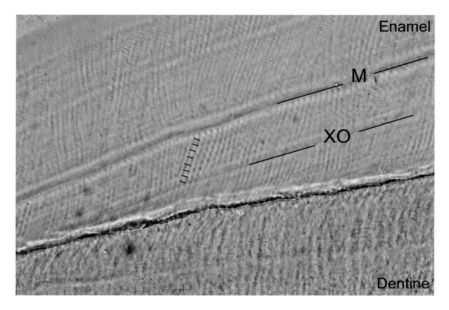

Figure 1.7
Experimental demonstration of daily lines in primate tooth enamel. A xylenol
orange (XO) label was administered and then followed 8 days later by a minocycline
(M) injection, and eight daily lines (red labels) appear between injections. Modified
and reprinted from Tanya M. Smith, "Experimental Determination of the Periodicity
of Incremental Features in Enamel," *Journal of Anatomy* 208, Issue 1 (January 2006):
99–113, with permission from Wiley-Liss. Macaque specimen courtesy of
Joyce Sirianni.

of this research, along with complementary studies by colleagues in England,
finally settled the debate about whether teeth have time lines.

It makes sense that teeth show daily rhythms, as many aspects of our
physiology cycle on a daily basis. Hormone levels, metabolic activity,
blood pH, body temperature, and sleep cycles all rise and fall over a 24-hour
period. The Japanese team hypothesized that daily lines in teeth are caused
by changes in carbon dioxide in the blood, which increases after falling asleep
and decreases during waking hours.[12] They believed that this cycle leads to
daily changes in the acid-base balance of the blood, causing daily variation
in the deposition of elements in teeth. Okada suggested that when blood
plasma became more acidic during the day, calcium levels in the blood
increased, which limited the precipitation of calcium in hard tissues—
producing a white, calcium-poor line. He reasoned that the opposite scenario

at night produced a dark, calcium-rich line. More recent studies have confirmed that there are subtle differences in the elemental composition and structure of daily features in enamel, leading to the characteristic light and dark banding pattern that is used to identify and count them.

It turns out that teeth also show rhythms that are both more and less frequent than the well-established 24-hour growth lines.[13] Experimental studies inspired by Okada and Mimura's work have proven that enamel and dentine also form lines every 8–12 hours. These closely spaced lines can be difficult to see clearly with a light microscope, and we don't fully understand why they form. Physiological rhythms that cycle more than once per day include our heart rate, body temperature, and hormone concentrations. Another team of Japanese researchers discovered that newborn rodents' teeth developed sub-daily rhythms prior to daily rhythms, suggesting a separate control mechanism for their formation. For the time being, our best guess is that hormonal or metabolic cycles that repeat throughout the day nudge the enamel- and dentine-forming cells to form minuscule lines in between the daily lines.[14]

Other mysterious structures repeat on a time scale greater than one day, and are thus referred to as long-period lines (figure 1.8).[15] Typically, their timing is determined by counting the daily lines between adjacent pairs of long-period lines in enamel. The most common rhythm in humans is 8 days, although different people may range from 6–12 days. Our working hypothesis is that the same long-period rhythm is found in the enamel and dentine of each permanent tooth from an individual, as well as in their bones. We're rather at a loss to explain what causes or controls the formation of these structures, as few aspects of our physiology cycle at this frequency. One folkloric explanation for the nearly weekly rhythm in human teeth is that these lines represent "feast days"—culturally prescribed days when people were likely to overindulge.[16] While consistent with Western cultural practices from nearly a century ago, we'll see in chapter 3 how there simply isn't any truth to this idea.

My colleagues and I have noticed that long-period rhythms in primates seem to relate to their body size.[17] Small primates show low values, starting from 2 days per line, while larger primates have a greater number of days between successive long-period lines. It is unclear whether there is a similar relationship in humans of varying size, since most counts come from people of unknown weight and height. Irrespective of their cause, determining the

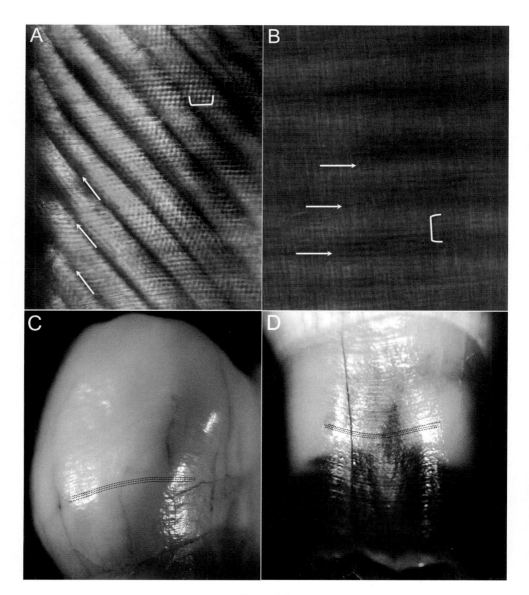

Figure 1.8
Microscopic biological rhythms in primate enamel and dentine. Long-period lines
inside teeth are indicated with white arrows in enamel (A) and dentine (B), and by
blue dotted lines on the outsides of the tooth crown (C) and root (D). Four daily
lines are indicated within each white bracket of A and B. Reprinted from Tanya M.
Smith and Paul Tafforeau, "New Visions of Dental Tissue Research," *Evolutionary
Anthropology* 17 (2008): 213–226, with permission from Wiley-Liss. Fossils courtesy
of the Natural History Museum (London) and the University of Liège (Liège).

long-period rhythm from a tooth is important, since it provides a short-hand way to estimate how long the tooth grew and how old young individuals were when they died. I'll demonstrate how this works in the following chapter.

Cementum, which grows much more slowly than enamel or dentine, records seasonal and annual rhythms in the roots of mammalian teeth.[18] A single yearly *cementum annulation* is identified as a paired light and dark layer (figure 1.6). The visual contrast is caused by variation in the mineralization and orientation of structural proteins within layers. Three mechanisms have been proposed for their appearance and timing, which posit seasonal variation in diet, seasonal variation in the hardness of foods consumed, or seasonal variation in hormone levels. An experimental study of goats by Daniel Lieberman provided support for the first two mechanisms. Animals on diets with varied nutritional quality and hardness showed cementum annulations, while those on a constant diet did not. However, this does not explain why humans living in industrialized societies show annual rhythms, as our diets don't vary that much from season to season. We'll circle back to this topic, because cementum annulations have potential for aging the skeletal remains of adults.

Uncovering the causes of biological rhythms in teeth is a kind of "holy grail" for my colleagues and me. It is a problem made all the more challenging by the fact that teeth are far removed from the rhythmic control center in the brain, receiving only indirect hormonal and neurological signals as they grow. Experiments during the past century have tested the influence of external cues on biological rhythms by keeping animals in constant light or dark conditions, as well as by subjecting them to temperature stress and changing feeding cycles. Anyone who's had extreme jet lag can attest that it's normal to feel out of sync while various organs catch up at differing rates to the new time zone! A few scientists have documented growth lines in hibernating animals, which undergo profound changes in their metabolism. The consensus of this research is that although environmental cues may shift the timing or diminish the appearance of incremental features in teeth, they are robust biological rhythms. In other words, the clocks that control their formation lie within the organisms themselves.

While the fine details of this system may only matter to specialists in skeletal biology and chronobiology, nature's biological rhythms aid other diverse fields of research.[19] For example, climate scientists count annual rhythms in trees to track environmental conditions backward in time for hundreds or thousands of years. Field biologists use yearly lines in bones and teeth to investigate growth, reproduction, migration patterns, and the age of wild animals. Paleontologists count tiny daily lines in fossil corals to document ancient climate variation and how the earth's position relative to the sun has shifted. And for biological anthropologists like me, rhythms in teeth provide a powerful tool for clocking the speed and duration of our growth, exploring developmental stress, and learning how old fossil children were when they met their untimely end.

Dental Tissue Studies Enter the Twenty-First Century

I've come to really appreciate how our knowledge of biology progresses, particularly as powerful investigative tools increase the pace of discoveries. Scientists began peering into microscopes to investigate tiny structures in teeth hundreds of years ago, beginning with Leeuwenhoeck's observations in the seventeenth century.[20] As standard light microscopes became widely available during the 1800s, European scholars elucidated our basic understanding of tooth structure. During this time, teeth were studied as "ground sections" under the microscope. This involved sawing, grinding, and polishing thin slices, which were then glued to microscope slides so that the fine internal details could be seen (figure 1.9).

Polarized light microscopy became popular in the 1860s, providing new information on the mineralized nature of teeth. This technique relies on optical filters that block light with certain wave properties, popular in modern anti-glare sunglasses. Polarized light creates dramatic contrast when it passes through complex crystalline arrangements, leading to striking colors and patterns in hard tissues and geological samples. Nearly a century later, focused electron beams were developed for imaging. These electron microscopes complemented information from regular light microscopy, further deepening our understanding of tooth development. As we've seen, this technique has illuminated the relationship between enamel-forming cells and their resultant prisms, as well as the size and shape of the crystallites that make up the prisms (figure 1.4). Another major advantage of electron microscopy is that it can reach much higher magnifications, even to the level of the atom.

Due to their durability and abundance, teeth have played a key role in defining the many species that have come before us. Yet because they must

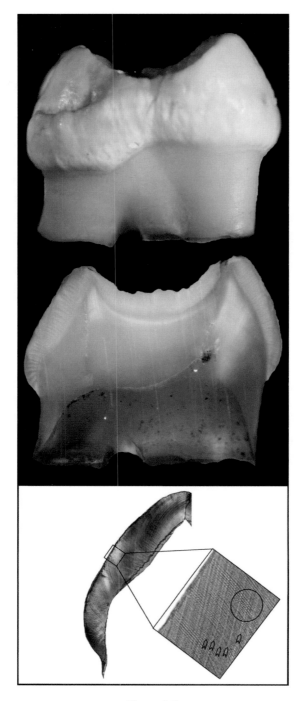

Figure 1.9
Molar before and after cutting to produce a section for microscopic imaging. Reprinted
from Tanya M. Smith and Paul Tafforeau, "New Visions of Dental Tissue Research,"
Evolutionary Anthropology 17 (2008): 213–226, with permission from Wiley-Liss.
Chimpanzee specimens courtesy of Christophe Boesch and Lawrence Martin.

be broken open, cut in half, or soaked in acid for in-depth developmental study, most fossil hominin teeth have been off-limits. Curators and custodians are charged with preserving rare material, and requests to cut fossils aren't often met with much enthusiasm. Fortunately, new X-ray imaging approaches have allowed us to peer inside hard tissues virtually. We can now unravel fine structural details and growth processes nondestructively.

Most people who have spent time in a dentist's office are familiar with radiography, the X-ray imaging method used to visualize the insides of our teeth and jaws. This works because dense structures such as bones and teeth stop some X-rays from passing through, and are easily distinguished in radiographic images from soft tissues that allow more X-rays to reach the radiation detector. Computed tomography—commonly referred to as CT scanning—relies on this principle. When a clinician calls for CT scanning to diagnose a medical issue, X-rays are focused on that part of the body from a number of different angles, and a digital camera captures these radiographs and feeds them into a computer. A software program then combines the radiographs to generate cross-sectional slices that approximate the body part as a virtual likeness. An important advantage of CT imaging over standard radiography is that this technique illuminates the body from different angles, including orientations that cannot be seen in standard radiographs, improving the accuracy of diagnoses.

Medical CT technology is good at imaging large objects quickly, such as unfortunate patients forced to lie still on narrow tables, but it doesn't work as well for very small things. In the last few decades, powerful micro-CT scanners have been developed to explore both large and small objects in great detail. Biological anthropologists have embraced this technology heartily, investigating the shapes of skulls, the internal design of the skeleton, and fine details of teeth while sitting at their computers (figure 1.10). Some use micro-CT imaging to locate important fossils in hardened blocks of soil in order to carefully remove them afterward, or to virtually rebuild skulls that have been broken or deformed during fossilization. Museums are also using micro-CT imaging to create digital databases of their collections, allowing objects to be shared and studied by scientists worldwide. As this technology becomes more affordable for research institutes and universities, the study of all kinds of fossilized life forms is taking a major step forward.

I've had a chance to witness this development firsthand. After finishing my PhD in New York, I moved to Leipzig, Germany, and headed up a Dental Hard Tissue Research Unit at the Max Planck Institute for

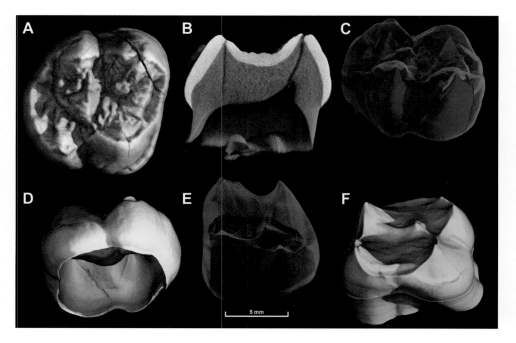

Figure 1.10
Micro-CT images revealing three-dimensional (3D) tooth structure. This is the
molar tooth shown in figure 1.9 prior to cutting. (A) 3D rendering of the chewing
surface; (B) cross-sectional slice through the front cusps; (C) 3D model of the tooth
showing the enamel cap (transparent yellow), enamel-dentine junction (red), and
dentine (transparent blue); (D) isolated enamel cap; (E) isolated enamel-dentine
junction; (F) isolated dentine under the enamel. Reprinted from Tanya M. Smith and
Paul Tafforeau, "New Visions of Dental Tissue Research," *Evolutionary Anthropology*
17 (2008): 213–226, with permission from Wiley-Liss. Chimpanzee specimen
courtesy of Christophe Boesch.

Evolutionary Anthropology. Our team was able to micro-CT scan thousands
of teeth over a few short years, demonstrating how this technology could
help distinguish fossil species. My career took another exciting twist when
I met Paul Tafforeau of the European Synchrotron Radiation Facility (ESRF).
Synchrotron imaging relies on a different illumination source than commercial X-ray generators. It uses fast-moving electrons that are bent by mag-
netic fields to produce synchrotron light. The wavelengths of these light
beams range from radio frequencies to high energy X-rays, which are determined by how fast the electrons are traveling. It turns out that seeing the
microscopic details of teeth requires a lot of energy!

Paul and I had completed our dissertations in the same year, coincidentally investigating similar questions about biological rhythms and dental development. After exchanging a few emails, he invited me to visit the ESRF for a preliminary experiment, which became the first of twenty scientific pilgrimages to Grenoble over the next decade. I became intimately familiar with the Lyon airport shuttle to and from the prestigious facility as we studied fossil teeth from all over the world. Synchrotron imaging is more powerful than medical or laboratory CT scanners, but the downside is that it is not as easy to use—or even to find. Most industrialized countries have one or two working synchrotron facilities, as they are expensive to build and maintain. Competition for access among the scientific and industrial communities can be intense. Our experiments would typically run night and day for 3–6 days, and it wasn't uncommon for one of us to nap in the control room between especially long scans. Eventually Paul figured out how to administer experiments remotely, allowing him to get a bit more sleep at home as long as someone was on hand to change samples on the beamline.

Scientists at the ESRF use a technique called phase contrast, which was originally developed to reveal subtle details that cannot be seen with regular microscopic illumination.[21] In his dissertation, Paul applied this synchrotron imaging approach to show tiny details of tooth structure. This simply wasn't possible with most micro-CT scanners, especially for dense objects like teeth. During our first meeting, he quickly convinced me that we could resolve growth lines without cutting samples open. We were excited by how this technique could revolutionize our field, since we had both come up against the hard truth that few curators would allow us to cut fossils for analysis. Paul and I were keen to validate the method, designing ways to compare it to standard micro-CT scans and slices of teeth prepared in my lab in Germany. Our first priority was to conclusively prove that daily growth lines could be seen in uncut teeth, which then opened the door for detailed studies of fossils.

We share a genuine passion for thinking about how teeth grow, and have spent long hours together poring over images, drawing tooth structures, and debating the interpretation of our results. When Paul and I started our research partnership, we knew that there were still gaps in knowledge of dental development, and had confidence that some of these might be addressed with synchrotron imaging. After working in Europe for a few years, I moved back to the United States to begin an Assistant Professorship

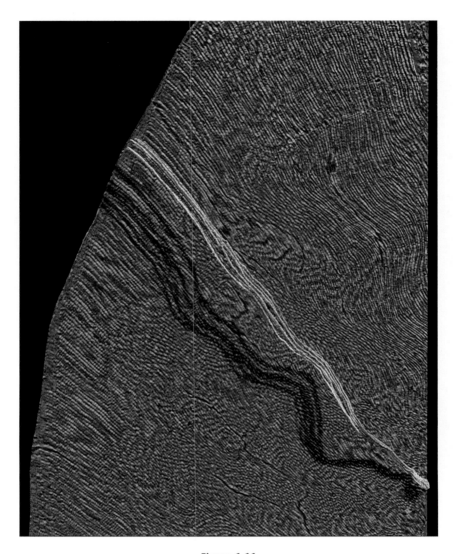

Figure 1.11
Synchrotron imaging model of enamel prisms in 3D above a virtual slice. (This
model can be seen animated here: www.youtube.com/watch?v=NMVbmTeE3xE.)
Enamel prisms run from the dentine (lower right) to the tooth surface (upper left).
Individual continuous prisms are represented by unique colors, which have been
traced with software. Daily lines can be seen as fine, paired light and dark bands in
the left side of the background. Modified from Paul Tafforeau, John P. Zermeno,
and Tanya M. Smith, "Tracking Cellular-level Enamel Growth and Structure in 4D
with Synchrotron Imaging," *Journal of Human Evolution* 62, no. 3 (2012): 424–428,
with permission from Elsevier. Chimpanzee specimen courtesy of Lawrence Martin.

at Harvard University, where I had the great fortune to mentor some amazing students. I quickly recruited John (JP) Zermeno, a gifted undergraduate who worked with us to model the three-dimensional growth of a tooth. JP spent months painstakingly tracing the outlines of enamel prisms on thousands of grainy synchrotron images. Together we learned that enamel-forming cells traveled complex paths during enamel secretion, which were not evident from other forms of microscopy (figure 1.11).[22] Perhaps more importantly, synchrotron X-ray imaging finally gave us access to some of those famous fossil juveniles discussed in the previous chapter.

I often joke with my students that after my "introductory" lectures on the development of teeth, they've learned more about this topic than if they had enrolled in dental school. Establishing this foundation allows us to begin working in my Dental Hard Tissue Laboratory. In a few short weeks they learn to create their own thin sections of teeth, from which they can identify these tiny structures themselves. I relish those moments of delight when someone first peers down a microscope at a tooth section they've made—recognizing features only seen in their books or my lecture material. Facilitating this experience for the next generation seems like a fitting tribute to the mentors who've patiently encouraged me to explore the microscopic world of teeth.

In the chapters that follow, we'll learn how scientists recognize important events during childhood, including our birth, the speed of our growth, and various illnesses and injuries. An appreciation of tooth development will also anchor our subsequent explorations of human evolution and behavior. For example, the emerging field of evolutionary developmental biology employs this knowledge to model the evolution of mammalian teeth. Scientists also investigate how far our ancestors roamed and what kind of foods they ate from elements in teeth. On an even broader level, aspects of tooth structure are relevant for engineers, who use biological designs to create resilient materials. Researchers in restorative dentistry are working to integrate the genetic and cellular details of tooth formation to grow replacement teeth. These promising areas of study owe a measure of their success to the early pioneers whose painstaking technical work and careful observations underlie our modern understanding of the natural world.

2   The Big Picture: Birth, Death, and Everything in Between

Gaining and loosing teeth are unavoidable experiences for most mammals—including humans. We've taken to ritualizing the loss of baby teeth with mythological figures such as the Tooth Fairy and the offering of monetary rewards. Yet there is more to these diminutive crowns than meets the eye. Our baby teeth and first molars have internal "biological birth certificates" that allow scientists to figure out the age of prehistoric children. Because our teeth grow and erupt into the mouth according to a predictable schedule, we can also use them to age living youngsters. This has borne out societally through oral examinations for military drafts, immigration, and employment verification. Tooth development finishes around age 20, when our third molars erupt and finish growing their roots—ending the addition of daily lines. As we progress through adulthood, our teeth age as well. They wear down, add tiny layers of cementum, and loosen their attachment to gum and bone. Understanding how teeth change over time makes it possible to age deceased adults and estimate the makeup of communities that were interred together. As you'll learn in the following pages, dental tissues hold important clues that span the life course, providing a unique perspective on life and death in the past.

The Birth Certificate in Your Mouth

It's pretty rare that you can point to a line inside a tooth and say: "I know where I was on this day"—even if you've been studying teeth professionally for years. But I got lucky. I had just set up my Dental Hard Tissue Laboratory at Harvard University when an old friend sent me the first baby tooth her son had lost, encouraging me to section and image it. After a few days of baking the small incisor crown into protective Plexiglas, I made two careful cuts with a rotating, diamond-tipped saw blade, removed a thin slice, glued it to a microscope slide overnight, and then ground and polished it until it was just 0.1 millimeters thick. I added a coverslip to the thin section and placed it on the stage of my new polarized light microscope. Once I turned on the light, set the magnification to 100 times, and peered into the eyepieces, I immediately noticed a dark line in the enamel near the cusp tip. This line separated a zone of clear, well-mineralized enamel from one that looked more variable and disturbed (figure 2.1). The dark line in Jimmy's tooth marked the day of his birth—a day I'll never forget, because I was on hand for his entry into the world. My friend had asked me to be her labor coach, and on that day I stayed close at hand for several hours, encouraging her to keep pushing until he finally appeared. The dark line in his tooth brought back the awe of that event almost 7 years earlier. I also felt a new sense of appreciation for the fact that we all carry tiny birth records in our mouths.

Baby teeth begin calcifying when we are still in the uterus, as does our first permanent molar (figure 2.2). These teeth preserve lines that indicate the temporary position of enamel- and dentine-forming cells during birth. Scientists discovered these *neonatal lines* by examining hundreds of baby teeth from human children, finding dark, accentuated lines in a similar position in most teeth.[1] Dental researcher Isaac Schour was the first to suggest that this microscopic disruption is caused by the physiological transition of birth. Others have since counted subsequent daily growth lines in young individuals, finding close agreement with their age and confirming that this line is indeed formed at birth. Under the microscope one can often see that enamel prisms deviate slightly from their straight course as they pass through the neonatal line. The adjacent enamel also shows a change in mineralization, producing a darkly contrasted line. Some hypothesize that these lines appear in infants' teeth because of changing calcium levels during or just after birth.[2] This is undoubtedly a stressful time for infants, who often lose weight for the first week or so after they are born.[3] Enamel secretion may even slow down immediately after birth, which researchers can detect from the compressed spacing of daily growth lines.

The neonatal line is quite broad in most individuals, indicating a disruption in tooth formation that lasts longer than one day. Scientists have wondered whether a difficult birth process leads to darker or broader neonatal lines. Current evidence appears to be mixed.[4] One comparison of 147 infants born by three different processes—naturally, via Cesarean section,

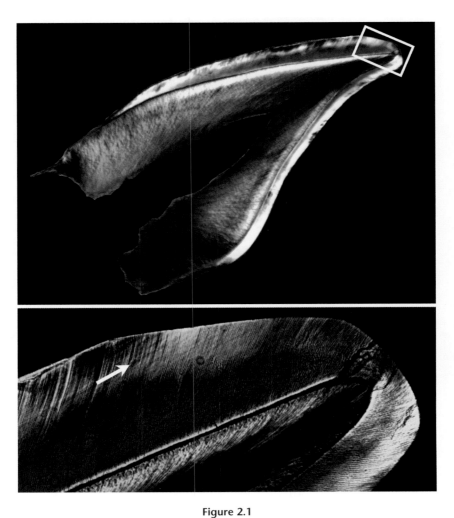

Figure 2.1
Naturally shed baby tooth showing the neonatal (birth) line. The upper image shows the enamel crown and remaining root dentine. A small white box reveals the area imaged at higher magnification in the panel below, where the neonatal line (white arrow) runs toward the worn surface of the tooth. Human tooth courtesy of Charis Ng.

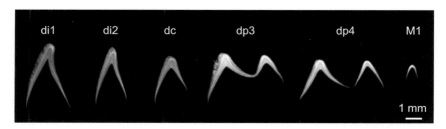

Figure 2.2
Hard tissue formation in the human dentition at birth. Virtual cross-sections gener-
ated from micro-CT imaging. Deciduous teeth are indicated by "d" and lowercase
letters—incisor (i), canine (c), and premolar (p)—and the permanent molar by a
capital letter (M1). Human jaw courtesy of the Peabody Museum of Archaeology
and Ethnology at Harvard University.

or with an intervention like forceps or vacuum—showed that the C-section
babies' teeth had the thinnest birth lines, followed by those born naturally.
Infants born after an operative intervention due to complications had the
thickest neonatal lines. However, another study of 100 children did not
find that these lines varied among children born different ways. My hunch
is that the discrepancies between these two studies may relate to the difficulty of preparing sections for microscopic analysis, which often involves a
combination of skill and luck.[5]

I've been curious about the sensitivity of infants to the birth process, and
how they experience their first few days and weeks in a strange new world,
which could help us to decode variation in ancient humans' teeth. It turns
out that nonhuman primate infants also have neonatal lines in their teeth,
as do other mammals. They appear in the first molars of free-living wild
primates, as well as in captive ones, and many show strongly marked birth
lines. This is interesting because nonhuman primates appear to have a more
mild labor and delivery process than humans. Their babies pass through
the birth canal with much less difficulty than large-brained human babies.
When one considers these various types of evidence, it doesn't seem possible
to determine *how* a modern human baby was born solely from the appearance
of the neonatal line.

However, this line does allow scientists to determine whether individuals survived the birth process. One example is an investigation of ritual-
ized human sacrifice in Carthage, an ancient Phoenician city in Northern
Africa.[6] The teeth of 50 young infants buried in ceramic urns were assessed

via microscopic study to determine if they were stillborn, or if they survived long enough to record growth after the neonatal line. Apparently half of the individuals did not preserve this marker, which means that they were unlikely to have lived long enough to be sacrificed. The research team concluded that, while occasional practices of this grisly ritual may have influenced historical reports, the prevalence of very young children in ancient Mediterranean funerary contexts is more likely due to stillbirths or infectious diseases during early infancy.

Another remarkable use of the neonatal line is the determination of whether an infant was born prematurely.[7] In one study, world-renowned histologist Christopher Dean teamed up with his former student Wendy Birch to analyze the baby teeth of a colleague's twins. Dean and Birch sectioned the teeth, located the neonatal line of each twin, and counted the number of days these teeth had been forming prior to birth. They came up with counts of 75 and 74 days for the two twins, which was 65–66 days shorter than the average prenatal formation time for third premolars. They inferred that the twins were born approximately 2 months prematurely. After completing their study, they learned from medical records that the twins were born 58 days before their predicted due date—confirming their conclusion. This approach could become a powerful forensic tool for investigations of unidentified remains from young individuals.

Incredibly, babies are not the only ones who form a line during birth—young mothers can too! We know this because the Japanese researchers who developed the time-marking system discussed in the previous chapter also examined the teeth of young female rabbits that had given birth.[8] Their study showed a broad light-colored area in the dentine a few days prior to delivery that switched to a dark band immediately afterward. They referred to this structure as a *parturition line* in keeping with the scientific term for giving birth, and hypothesized that it is caused by changes in the carbon dioxide and acidity of blood. Subsequent experimenters have confirmed that a parturition line may be seen in other mammals, including dolphins and ground squirrels.

The exciting possibility of finding a similar line in human teeth motivated Christopher Dean to team up with Fadil Elamin, a Sudanese dentist. They examined the teeth of Sudanese women who had given birth while they were teenagers, prior to the completion of their third molars around 20 years of age. Recall that after this point, teeth stop adding daily lines, thus the importance of choosing women who became mothers at young ages. Dean and Elamin found several broad lines in the third molars of four women. These resembled the parturition lines found in other mammals, and nearly half of the lines were believed to match with birth events. Their results show potential, yet the study also underscores how complicated it is to be definitive about their identification in humans. While first molars show neonatal lines in a consistent location, allowing assessments of age, third molars begin forming several years after birth, making it more difficult to tie their formation to a precise age. Moreover, memory tests prove that most people are not able to recall past events accurately. It isn't clear how well the Sudanese mothers estimated the timing of their deliveries years later, since they couldn't be verified with hospital records. And there is a final complication— it's nearly impossible to rule out alternative causes for similar lines that form after the neonatal line. In the following chapter I'll review a number of other potential disruptors of tooth formation that can resemble parturition lines.

Similar complications also plagued the first report of a parturition line in a captive macaque monkey.[9] Jacqui Bowman observed a line in the third molar of a young female that occurred near the age that she gave birth to a stillborn infant. However, the mother was unwell around this time, requiring anesthesia and dying shortly after delivering the infant. Thus it was difficult for Bowman to rule out that illness or medical intervention caused the accentuated growth lines rather than the experience of giving birth. My colleagues and I met a similar difficulty during an analysis of another captive female monkey.[10] This individual showed numerous accentuated lines in the final 6 months of her life, during which she was hospitalized several times. She also gave birth to a stillborn infant a mere 50 days after its conception. Unfortunately I couldn't be sure that any of the numerous accentuated lines were a secure match to the birth of the infant or its conception.

While these studies may seem cruel at first glance, researchers did not harm the monkeys to cause disturbances in their teeth. These were a by-product of their illnesses, medical treatment, and possibly the delivery of their infants. I've included them here as they demonstrate the importance of using medical records when they are available to interpret events in the teeth of individuals, a subject we will consider further in the following chapter. Thus far we've been unable to reconstruct the reproductive histories of slow-growing species like humans or nonhuman primates from the lines in their teeth alone. Having a reliable way to pinpoint childbirth could help to address unresolved questions about the evolution of human reproduction. As we'll discuss in later chapters, combining analyses of elements in teeth with the timing of their formation may provide a way forward.

Behold the Tooth Fairy!

Parents of young children can attest that teething is often the next uncomfortable rite of passage for infants following birth. Babies' gums become swollen and stretched as tiny white bumps and ridges poke out during the first few months and years of life. When teeth emerge through the gum, they begin the visible part of tooth eruption, a long journey from the inside of the jawbone to their final position in the mouth (see figure 1.3). Baby teeth are sometimes called milk teeth, since they erupt while children are nursing, and all 20 teeth are in place by about 2.5 years of age. These teeth are smaller and fewer in number than the permanent adult dentition. Just a few years after the milk teeth appear, they are shed like the leaves of deciduous trees, and thus they are also called *deciduous teeth*.

At age five or six, children start to lose their tiny deciduous incisors through impatient wiggling of loose, unfamiliar-feeling crowns, or perhaps unexpectedly while biting into something. This often marks our first conscious experience of the impermanence of our teeth. The legend of the Tooth Fairy may prompt a monetary gift in exchange for the precious tooth. The Tooth Fairy appears to have become pretty generous in the past few years—American children received an average of $3.70 per tooth as of 2013. A small gift or monetary reward is thought to soothe the feeling of loss, while introducing the cultural norm of currency exchange. The American version of this legend dates to the beginning of the twentieth century, although rituals marking the loss of the first tooth predate this particular custom.[11] In a number of cultures an animal such as a mouse, rat, or crow is symbolically tasked with the replacement of lost teeth. Certain European practices once included having a parent or child swallow the tooth in order to prevent future toothaches. Parents in the United States seemingly prefer to shell out a few dollars and pocket the tooth instead!

The truth is that our deciduous teeth don't randomly fall out. Special cells begin to break down their small roots as permanent teeth begin growing underneath. If you look closely at the roots of naturally shed baby teeth, you'll notice that often only a small point or ring of dentine remains. This loss creates space for the next generation of incisors, canines, and premolars—which then begin their advance into the mouth. Our permanent molars are the only type of tooth that doesn't have a deciduous precursor, which makes sense, since there isn't any room for them in the tiny jaws of infants. As our faces and jaws elongate and broaden during adolescence, additional space becomes available to house these molars, typically bringing the full complement of permanent teeth to 32.

Tooth eruption occurs at fairly predictable ages in modern humans, except for third molars, which are sometimes missing or stuck in the jawbone (table 2.1). By about age 6, the first molars and incisors appear, generally followed by the canines, premolars, and second molars by age 12. Our third molars may make their appearance starting around age 18, earning them the name wisdom teeth as the dental marker of adulthood. The term comes from an ancient Greek phrase, which seems fitting given the philosophical inclination of the Greeks.[13] Aristotle's writings on natural history mention them, including the fact that they could sometimes appear in very old individuals and cause great discomfort when they failed to erupt properly. This condition of chronic tooth impaction must have been awful to live with in

Table 2.1
Average Eruption Ages of Human Teeth Past the Gum Line (Gingival Eruption)[12]

Tooth	Age, upper jaw	Age, lower jaw
di1	10.0 months	8.0 months
di2	11.4 months	1.1 years
dc	1.6 years	1.7 years
dp3	1.3 years	1.3 years
dp4	2.4 years	2.2 years
I1	6.8 years	6.2 years
I2	8.0 years	7.0 years
C	11.4 years	9.8 years
P3	9.9 years	10.0 years
P4	10.8 years	10.6 years
M1	6.4 years	6.3 years
M2	12.6 years	11.8 years
M3	16.5 <> 20.5 years	16.5 <> 20.5 years

Deciduous (baby) teeth are indicated by "d" and lowercase letters—incisor (i), canine (c), and premolar (p)—and permanent teeth by "M" and capital letters (I, C, P). Ages averaged for males and females of European ancestry.

the days before oral surgeons. In chapter 3, I'll explain how it has become more commonplace as our diets have changed over the past 10,000 years—although happily, many people have access to much-improved dental care when molar impaction does strike.

Scientists don't fully understand how teeth move through the jaw during eruption. Pressure from the lengthening tooth roots, the periodontal ligament, and the soft tissues underneath the growing tooth certainly play a role.[14] When teeth pass beyond the margin of the jawbone they are said to have gone through *alveolar eruption*, which can be seen in living individuals with radiographic imaging (figure 2.3). In order for the tooth to pass into the mouth, the bone covering the tooth is broken down and resorbed by special cells. This process creates a hole in the jawbone for it to pass through, which can be recognized in skeletal remains. A number of young fossil hominins died while their teeth were making their way through the jaws. As Raymond Dart noted for the Taung Child, key events like dental eruption can be compared among individuals to infer their age. This helps us to benchmark their rate of growth, as we'll see in chapter 6.

After passing through temporary windows in the jawbone, teeth pierce the gum in a process known as *gingival eruption*, which can take several months. This is the part that babies and their caretakers would like to happen faster! Alveolar eruption is well underway when teeth begin to cut the gums (figure 2.4). As you can see in this comparison of a monkey tooth, approximately one-half of the tooth has emerged past the bone, yet only two molar cusp tips have passed the gum line. It is difficult to estimate the timing of gingival eruption from ancient remains—during life, the gums cover teeth with several millimeters of tissue that decomposes quickly after death. This is important because we're often forced to compare teeth in skeletonized jaws with information from living subjects with healthy gums. Several months are believed to pass between alveolar and gingival eruption, which likely varies in different tooth types and among primates—complicating comparisons of deceased and living subjects.

When the tooth crown reaches the level of other erupted teeth in the mouth, it encounters its opposing upper or lower tooth. Thus begins the lifelong process of tooth wear, which could serve to shorten the tooth after a significant amount of use. The body has come up with an elegant solution to maintain tooth-tooth contact over time—it continues the process of eruption at a much slower rate. Tension in the ligament that holds the

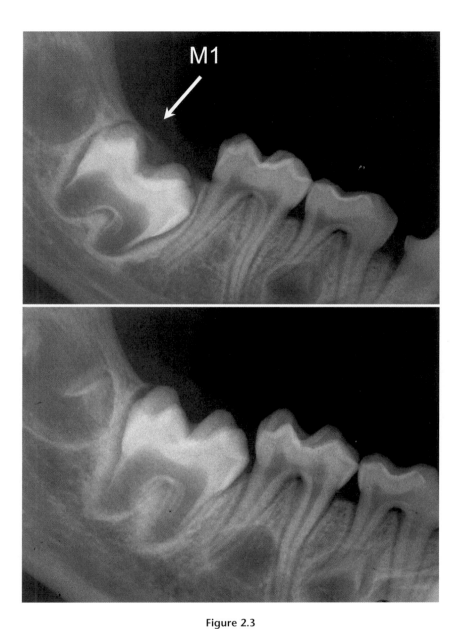

Figure 2.3
Alveolar (bony) emergence of the first molar of a baboon. The permanent first
molar (M1) was radiographed just before emergence (above) and after passing
through the bone three months later (below). Reprinted from Jay Kelley and Tanya M.
Smith, "Age at First Molar Emergence in Early Miocene *Afropithecus turkanensis* and
Life-history Evolution in the Hominoidea," *Journal of Human Evolution* 44 (2003):
307–329, with permission from Elsevier.

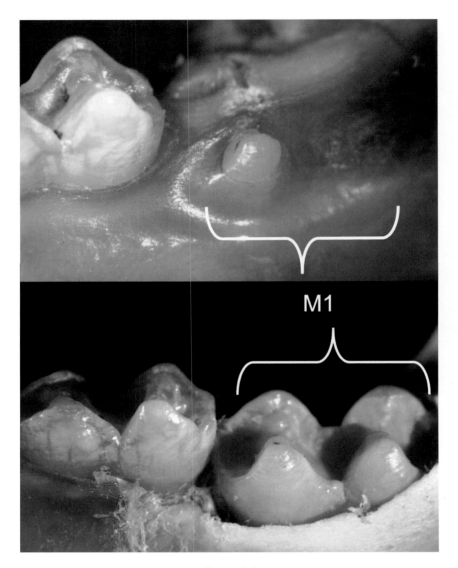

Figure 2.4
Gingival (gum line) emergence of the first molar of a macaque. This specimen was preserved in ethanol after death, allowing a comparison of the first molar (M1) relative to the gum line (above) and bone after dissection (below). Image credit: Akiko Kato and Tanya M. Smith.

tooth in its socket keeps it in contact with the opposing tooth, which is reinforced through gradual cementum deposits at the base of the root.[15] This continuous eruption was particularly important in the past, as early hominins and modern humans who relied on wild foods wore down their teeth more quickly than those of us who rely on soft, industrially processed foods. While this may seem like a good thing, our new diets have also led to a number of dental problems that are the subject of the following chapter.

I Don't Wanna Grow Up: Age Assessment in Children

One of the first things people often ask young kids when meeting them is how old they are. Age provides context for how someone sees the world, such as belief in the Tooth Fairy, and especially about how much longer they'll be growing. This latter point has been of keen interest for biological anthropologists. Debates about when a human-like growth pattern evolved in our ancestors are almost as old as the field of anthropology. Unfortunately, we can't ask fossil hominin children how old they are. But we can query their teeth for this information.

Tooth growth is more protected from typical fluctuations in nutrition and health during childhood than other aspects of skeletal development—although it is not invulnerable. This means that dental development is a more accurate predictor of a human child's age than other measures like height, weight, or the size of bones. Our teeth mineralize in a fairly predictable sequence and at reasonably consistent ages. Recall that the process begins as the enamel- and dentine-forming cells build tiny cones of hard tissue. These enlarge into characteristic crown shapes, followed by the secretion and mineralization of roots. In order to know how many teeth have begun to calcify, and how well developed they are, we can examine dental radiographs (figure 2.5). These images reveal the status of teeth that have yet to emerge into the oral cavity, as well as the root formation of those that have erupted—details that are not evident from simply glancing into someone's mouth or looking at a jawbone.

Growth standards have been created from children in Western countries, where clinical radiography has been used for decades. This involves categorizing the calcification stage of each tooth for many children at different ages, yielding a data set that can predict a child's age from their own stages.[16] For example, the youngster shown in the top of figure 2.5 has

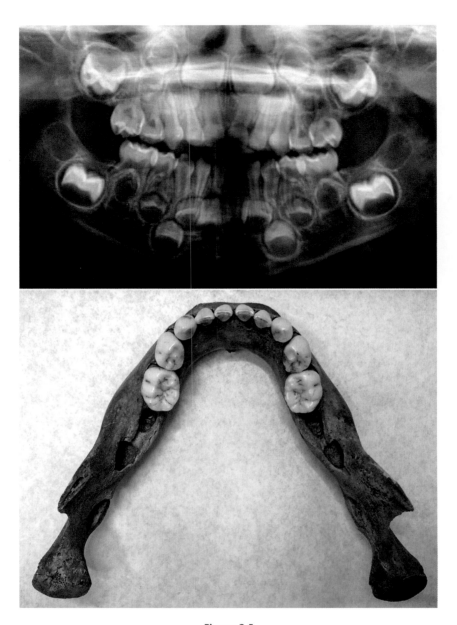

Figure 2.5
Radiographic and visual inspection of tooth development in two similarly aged
human children. Radiograph courtesy of Masrour Makaremi. Human jaw courtesy of
the Phoebe A. Hearst Museum of Anthropology and the Regents of the University
of California (catalogue no. 12-10403(0)).

several permanent teeth forming in the bone above and below erupted deciduous precursors, and the outlines of the adjacent "M-shaped" lower first molar crowns are almost complete, which happens around 3 years of age. The actual age of this boy is 2 years and 10 months.

Another approach to age prediction is a simple visual inspection of the mouth of living subjects or the jawbones of skeletonized ones. In this case, one notes which teeth have erupted, which have yet to erupt, and whether any teeth are in the process of erupting. For example the lower jaw shown in the bottom of figure 2.5 has erupted each deciduous tooth, but no permanent ones. This information is then compared to average eruption ages for a reference population (table 2.1). In this example, the individual is likely to be older than 2.2 years old, the age that the final lower deciduous premolars erupt, and younger than 6.2 years, as the permanent first incisors and first molars are not yet visible. While this is easier than obtaining and scoring radiographs, it is less precise than evaluating how well mineralized each tooth is. It isn't possible to know if a tooth has failed to erupt because it never formed, is still growing, or has become impacted. As we've seen, impaction is a common problem with the third molars, making them less reliable for age determination unless one uses radiography to see what's going on under the surface.

Pediatric dentists have spent many years studying tooth mineralization and eruption, creating helpful tables and charts that can be used to age dentally immature individuals (figure 2.6). Comparisons of estimated ages with actual ones show that these growth standards can be accurate to within a year of an individual's age, particularly for children under age 15.[17] In the case of our young boy, the information in figure 2.6 would suggest that he was between 2.5–3.5 years old, a range that includes his true age.

A troubling example of the use of aging standards dates from the "Industrial Revolution" of the 1800s.[18] During this period, the economies of Western countries shifted rapidly from agricultural production to machine-based industrial production. Young children were often forced to work in factories, mills, and coal mines. Eventually, the British parliament passed a law that children had to be at least 9 years old to work in certain industries. However, because birth certificates were not common, ages were often subjectively assessed—and easily manipulated. In 1837, following a study of dental development in more than 1,000 children, a British dentist proposed

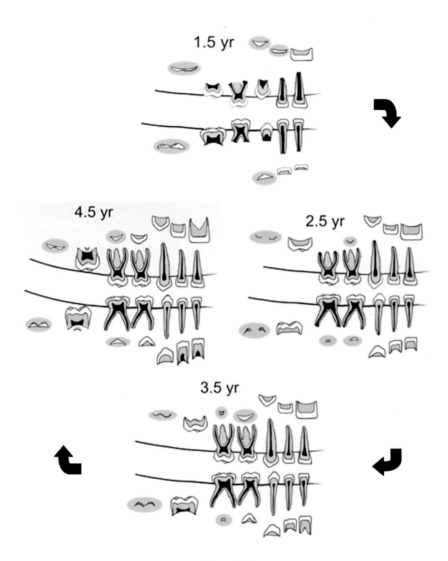

Figure 2.6

Schematic of human tooth calcification and eruption over time. The deciduous dentition is in grayscale, and the permanent dentition is in teal. The calcification status of each tooth is shown from a radiographic perspective, and the eruption status of each tooth is shown relative to horizontal lines that signify the position of the jawbones. Modified with permission from the full atlas of S. J. AlQahtani, M. P. Hector, and H. M. Liversidge, "Brief Communication: The London Atlas of Human Tooth Development and Eruption," *American Journal of Physical Anthropology* 142 (2010): 481–490.

that tooth eruption could be used to determine whether children were old enough to work legally, a practice that was quickly adopted.

A similar use of child labor is prevalent today in India. Although the law states that children must be at least 14 years old to work, exploitation of underage children is common, particularly since many children are not registered at birth.[19] Advocates are beginning to press for laws to ensure that dental eruption standards are used for aging for Indian children. One can hope that this information will empower human rights advocates to better protect children worldwide.

Being able to age juveniles has been important in social contexts since at least the days of the Roman Empire.[20] At that time, male children erupting the second molars around age 12 were considered old enough for military service. Contemporary authorities often have to assess whether undocumented foreign nationals, such as political asylum seekers, should be regarded as adolescents or adults. A writer from *Slate* magazine once contacted me to ask if it would be possible to determine if a captured Somali pirate was still a juvenile. I explained that a simple radiograph could tell us about the development of his third molars. Since they are typically incomplete in adolescents, this could help to approximate his age. However, while radiography and visual inspection can establish if someone has yet to "grow up," sound ethical reasons prohibit approaches that would yield a more precise age for a living person. We can't, for instance, use standard microscopic imaging because this requires removing and sectioning healthy teeth, nor should we employ a lethal dose of synchrotron X-rays to reveal growth lines. Thus, dental X-rays are the most appropriate technique for aging living children.

When scholars are tasked with studying dental remains from deceased juveniles, we can use their faithful biological rhythms to arrive at more accurate chronological ages. Recall that cells record the passing of days as tooth crowns and roots take shape. To read this microscopic map and calculate an individual's age, we must determine the time taken to form the enamel and dentine secreted after the neonatal line.[21] This could become incredibly tedious, particularly for older children, since it requires counting thousands of tiny lines! The good news is that there are a few shortcuts that help us avoid having to count every single day. I'll take you through the logic here, which involves some basic math. To estimate the time taken to form the first part of the crown, we divide the distance enamel-forming cells travel during secretion (prism length) by the daily secretion rate. Keep in mind the mathematical definition of a rate, which is distance divided by time. Because each pair of light and dark lines is formed in 24 hours, the secretion rate is simply their average spacing divided by one day. With a microscope we can easily magnify and measure the length of daily lines in enamel (figure 2.7). To determine the formation time of the rest of the tooth, we count the long-period lines that come to the surface of the tooth,

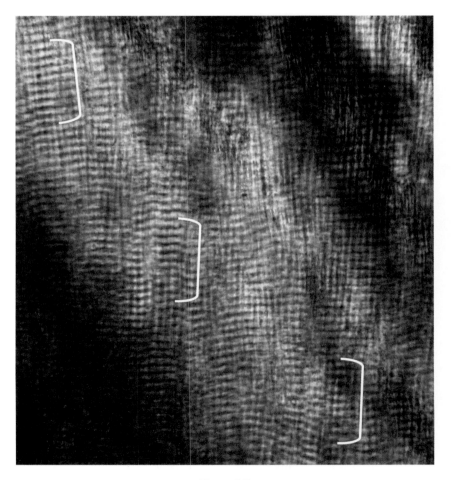

Figure 2.7
Daily growth lines in a fossil hominin tooth. White brackets indicate groups of 10 daily lines (paired light and dark bands) spaced, on average, 5.5 microns apart. Fossil courtesy of the National Museum of Ethiopia (Addis Ababa).

and multiply them by the number of days between an adjacent pair of these lines.[22] Adding together these two shortcuts gives us the time it takes to form an entire anterior tooth, or the cusp of a premolar or molar tooth.

This might make more sense with an actual case study. A few years ago, I mapped the growth of a molar from a wild chimpanzee who was the first confirmed lethal case of Ebola in this species.[23] Removing and sectioning the tooth was a little unnerving, but I had been reassured that the skeleton had been thoroughly decontaminated prior to its export from the Ivory Coast. After creating a thin section in my lab, I found the neonatal line under high magnification and determined that it formed 31 days after the molar cusp had begun calcifying (figure 2.8).

I then measured the rate of daily secretion in the enamel, and divided the length of a central enamel prism by this rate to estimate that there were 166 days of growth between the neonatal line and the cusp tip. From there I counted 105 long-period growth lines in the enamel formed after the tip of the cusp, which I multiplied by 5 days—the number of daily lines between long-period lines—to get 525 days. I then added 166 and 525 to determine that this part of the crown finished forming at 691 days of age.[24] To estimate root formation time, I counted long-period lines in the dentine, estimating that there were 141, which I multiplied by the 5-day repeat interval to determine that the root had been forming for 705 days before the animal died. By adding 691 days of postnatal crown growth to 705 days of root growth, I estimated that this chimpanzee died at 1,396 days of age, or 3.82 years. When my colleague checked his field notes, he reported that the animal likely died at 1,372 days of age, which is a difference of only 24 days—a mere 2% error!

Similar analyses of captive animals have reported even better accuracy, and studies of recent human remains also produce reliable results.[25] As a professional stickler for detail, I should confess that there are cases in which this doesn't work. The teeth must be well preserved after death, and the thin sections need to be cut properly. This method requires specialized equipment, considerable experience, and a good deal of time. Difficult cases in which the teeth are hard to interpret can take weeks to analyze, and there are instances when it's simply impossible to age an individual. These challenges may explain why forensic experts do not use this approach more often, despite its value for aging the remains of young individuals. Only a few forensic studies of this nature have been published, including one where the remains of an infant were assessed from tooth growth lines, historical

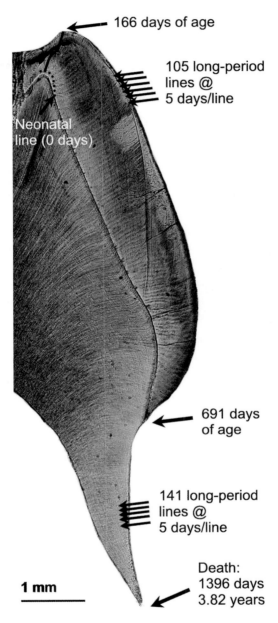

166 days of age

105 long-period lines @ 5 days/line

Neonatal line (0 days)

691 days of age

141 long-period lines @ 5 days/line

1 mm

Death: 1396 days 3.82 years

Figure 2.8
Reconstruction of age at death from a first molar cusp. Chimpanzee specimen courtesy of Christophe Boesch.

records, and DNA analysis.[26] Growth-line analysis indicated that this child was between 4.8–5.1 months old, consistent with the historical identification of a 5-month-old female. The main use of this time-consuming method has been to age young nonhuman primates and fossil hominins. Microscopic study is particularly valuable in these instances, since it isn't possible to apply human tooth calcification or eruption standards to arrive at accurate ages, as different species grow up according to their own schedules.

Keeping It Real: How Old Are You, Grandpa?

While familiarity with dental development is invaluable for determining the age of a youngster, I'm often asked whether it is possible to figure out how old adults are from their teeth. This is much more challenging than aging children, but no less important. For anthropologists who excavate burial mounds or ancient cemeteries, determining how long people lived and how many senior individuals were members of a social group opens important windows into the past. Forensic scientists typically assess skeletal features to determine adult age—such as the shape of joints in the pelvis— but these methods are often imprecise and inaccurate.[27] Scholars in my field have struggled to develop more effective approaches for decades.

It's fair to say that teeth "age," just as other components of our bodies age over the years, yet the ways adult teeth change are less predictable than the growth of juvenile teeth. One example is modification through wear, also called *attrition*. This is easy to see but challenging to measure. As tooth cusps encounter food and the opposing teeth during chewing, the most-used areas begin to show distinct points of contact. With continued use these contact facets enlarge and deepen, exposing the underlying dentine. Anthropologists often make a casual diagnosis of whether an individual is an adult through a quick inspection of the number of teeth erupted and the degree of wear on the late-erupting teeth. For example, a human jaw with a full set of erupted teeth and wear facets on the third molars is most certainly an adult. But can one determine how old this individual is?

The British dental scholar Albert Edward William "Loma" Miles proposed an innovative method to estimate an adult's age built on an understanding of tooth development. Knowing that gingival eruption occurs at predictable ages allowed Miles to estimate the timing and rate of molar wear. Recall that first molars erupt at 6 years of age, second molars at 12 years, and third molars at about 18 years on average. By looking at a series of individuals erupting their second molars, Miles was able to estimate how much a first molar typically wears down over 6 years, which is the time between when it erupted (age 6) and when the second molar was erupting (age 12). He performed a similar comparison of second molars in individuals erupting their third molar, giving him a sense of what 6 years of wear on the second molar looked like. Miles was then able to extrapolate for the third molars based on the patterns already established for the first and second molars. For example, a third molar that shows as much wear as a first molar at 12 years of age has been in use for 6 years. Thus this individual is close to 24 years old, which is the average age the third molar erupts (18 years) plus the estimated duration of wear (6 years). A test of this method on individuals of known age revealed that it could be fairly accurate, with estimates deviating from actual ages by a few years or less.[28] As you might guess, this method works best for young adults. We can't predict the age of individuals who are older than about 50 with nearly as much confidence.

An alternative approach is based on changes in the structure of the root dentine (figure 2.9).[29] Remember that long, thin channels perforate newly formed dentine. Over time, these channels gradually fill in with mineral and protein as the tail-like cellular projections retreat to the pulp space, where bodies of the dentine-forming cells reside. When these channels fill in, the dentine becomes more uniform, allowing light to pass through more easily. Freshly formed dentine tends to scatter light as it passes through the channels, making it possible to see what look like long filaments running through the opaque dentine. In contrast, areas in aged dentine with filled channels are like transparent clear glass. An example of this can be seen in the lower part of the chimpanzee root in figure 1.6. This translucency spreads from the bottom of the root upward toward the crown over time. Scientists measure root translucency by shining light through intact teeth or thin sections, and can predict adult ages to within 4–7 years.

Additional changes that occur in older people include the infilling of the pulp space by the dentine-forming cells, which deposit tiny amounts of dentine continuously throughout adulthood.[30] This process also shows a relationship with age, although it is not as effective for age prediction as measurements of root translucency. The cementum holding the roots in place also increases in thickness over time, and this trend has been used for age prediction, although this is even less accurate than the other approaches.

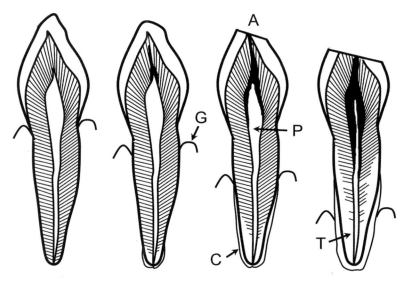

Figure 2.9
Methods to assess adult age from teeth. Progressive stages of aging are shown from left to right, which include gum recession (G), attrition (A), pulp space infilling (P), increasing cementum thickness (C), and dentine root translucency (T). Redrawn from G. Johanson, "Age Determination from Human Teeth," *Odontologisk Revy* 22, Suppl. 21 (1971): 1–126.

Cementum can thicken in response to direct pressure, as from chewing, but it is also commonly removed in other places. So measurements of its thickness can be confusing.

A more obvious age-related change is the height of the gum line, which recedes with age. This explains the phrase "long in the tooth," since roots become more visible due to gum recession—thus our teeth appear to get longer. Measurements of recession alone are not very reliable for age prediction though, since chronic gum inflammation or periodontal disease can cause premature and uneven gum deterioration, a common problem for humans living today.[31] Scholars have attempted to combine approaches to increase the accuracy of age predictions, and this is more effective than using any one technique. An important caveat is the fact that diet, oral health, genetics, and nondietary uses impact the structure and appearance of teeth, and controlling for all these influences when determining an individual's age is nearly impossible.

A final method for aging adults is based on counting cementum lines, which are formed yearly once teeth erupt into the mouth. This technique can be effective for aging mammals, particularly from seasonal environments, but it has met with mixed success in living humans.[32] One influential study examined cementum annulations in teeth from 80 known-aged individuals.[33] Of these, 73 were deemed suitable for analysis, and their predicted ages were within 6 years of their true age on average. These initial results inspired others, including anthropologist Ursula Wittwer-Backofen, who conducted an ambitious study of 433 human teeth of known age. Her team employed chemical treatments and image analysis software to improve the accuracy of their counts.[34] The results from 363 teeth were quite promising, with the majority of calculated ages falling within 3 years of the actual age. However, more than 15% of the teeth could not be counted because of the poor quality of the resulting images. She and her colleagues concluded that—given the possibility of erroneous counts—at least two teeth should be assessed for each individual, and counts should be combined with root translucency analysis to arrive at a more secure estimate.

One concern about this method is that counts of the same teeth by different observers may vary by as much as 8 years, meaning that it may not be more accurate than other less laborious aging approaches.[35] The accomplished dental anthropologist Simon Hillson notes that what initially seemed like a very promising approach might be a little too good to be true for long-lived humans. Cementum annulations are very closely spaced, making them harder to resolve with light microscopy than rhythmic features that form more rapidly. Of bigger concern is the fact that cementum isn't deposited evenly or at a constant rate, leading to variation in different areas along the root surface. "Some teeth produce highly anomalous counts," Hillson notes,[36] echoing my own experience of trying to document these pesky yearly lines in known-aged humans and great apes.

Despite these mixed results, tooth cementum analysis continues to receive considerable attention from biological anthropologists, archeologists, and wildlife biologists.[37] Ongoing research aimed at optimizing this method through formal analytical protocols, software that enhances image quality, and the use of nondestructive synchrotron imaging will hopefully improve things. Strengthening our methods for aging adults could tell us more than how long they lived—it may help us estimate how many children ancient hominins may have raised, when their lifespans elongated, and how their family groups were structured. Until then we must rely on other kinds of dental evidence for the evolution of human development, the subject of chapter 6.

We've now covered how tiny growth lines allow us to determine when someone was born, how fast their teeth grew, and the age of those who didn't survive childhood. This type of painstaking analysis is only one of several methods to determine age; other approaches employ large numbers of children to characterize the timing of tooth calcification and eruption. Predictive charts help forensic scientists and anthropologists age skeletal remains, although these standards are largely based on children from industrialized countries. As you'll learn in the following chapter, factors such as nutrition and illness can impact the formation of teeth. Of greater concern for those of us who estimate age from these charts is the possibility that human populations differ in the timing of tooth formation. A "one-size-fits-all" model of tooth formation might be an oversimplification. Ongoing research on humans from across the globe is likely to further improve age estimates. Radiographic imaging and visual inspections complement in-depth studies of tooth microstructure—and next we'll explore how this latter method can reveal even more intimate details of an individual's history.

3 Things That Can Go Wrong: Stress, Pathology, and Dysevolution

In the previous chapters we considered how teeth take shape, increase in size, and emerge into the mouth, only to begin wearing down and changing with age. Over the course of crown and root development, bodily disturbances may lead to microscopic defects or—less commonly—teeth may fail to form or erupt properly. Once our teeth do emerge, oral bacteria and dietary factors contribute to cavities and gum inflammation, which may ultimately lead to tooth loss. Some of these experiences have plagued us since ancient times, while others are believed to be diseases of modern civilization. The invention of agriculture and eventual production of food on an industrial scale have profoundly affected our physiology and anatomy, including our teeth. In this chapter, we'll explore developmental, evolutionary, and behavioral perspectives on dental disruptions and diseases.

Childhood Illness and Nutrition Writ Small

The milestones of birth, tooth emergence, and the loss of deciduous teeth encompass an incredibly vulnerable time for human children. Born toothless, during the first few months of life babies rely exclusively on their mothers for nutrition. In small-scale traditional societies, soft nursing foods are introduced around 6 months, close to the age that infants' first teeth emerge.[1] For the next few years it seems that everything goes into the mouth for a taste or a nibble! Babies aren't just learning about food during this time, they're also encountering new immunological challenges. Their digestive and immune systems are developing rapidly, and bouts of gastric distress and illness are unavoidable. During the transition from breast milk to an adult diet, new foods test these systems further, particularly in countries where clean water and sewage disposal are limited.[2]

Most permanent tooth crowns calcify inside the jawbones between birth and age six, serving as powerful windows into the health and well-being of young children. As we've seen, teeth record much more than the daily rhythms of life. Profound changes during birth indelibly mark the teeth of infants, as do subsequent illnesses and injuries. Major disruptions can produce *hypoplasias*, areas of depressed or missing enamel or dentine where the cells stop secreting prematurely—causing a pit, plane, or ring around the tooth (figure 3.1). More subtle disturbances appear internally, which we generically call *accentuated lines*. These alterations in enamel or dentine structure can be used to age a disruptor, since they permanently capture the position of cells that were actively secreting at that time (figure 3.2). In the preceding chapter we learned about two specific types of accentuated lines formed at birth in the teeth of infants (neonatal lines) and mothers (parturition lines).

Peering through a microscope at the teeth of prehistoric humans reveals that childhood has always been stressful, and probably even more so in the past.[3] Many anthropologists have documented hypoplasias and accentuated lines in fossil hominins, great apes, and preindustrial modern humans. So what can these seemingly ubiquitous features tell us about childhood? Many things—indeed, perhaps too many things, since they may be caused by numerous conditions (table 3.1). We can organize these into several broad categories: specific illnesses or symptoms; nutritional deficiencies; and physical or psychological stresses. Many causes have been deduced from experiments on laboratory animals or observations of captive primates with associated medical records. Careful studies of humans with developmental defects and documented histories are quite rare, but they make for illuminating reading.

In the previous chapter, we discussed an investigation of human twins in which their premature birth was revealed by unusually low numbers of daily lines formed before birth.[4] The authors of this study also counted days after the twins' birth to age a series of accentuated lines. They compared their estimates to medical records and found that approximately one-third of the accentuated lines could be related to an event. For example, a near-perfect match was obtained on the day that three routine vaccinations were administered to both twins. A similar correspondence was found when the twins were transferred out of intensive care and when one twin was discharged from the hospital, as well as when the other was readmitted for a surgical procedure. Gastrointestinal upset, eye problems, and vomiting all

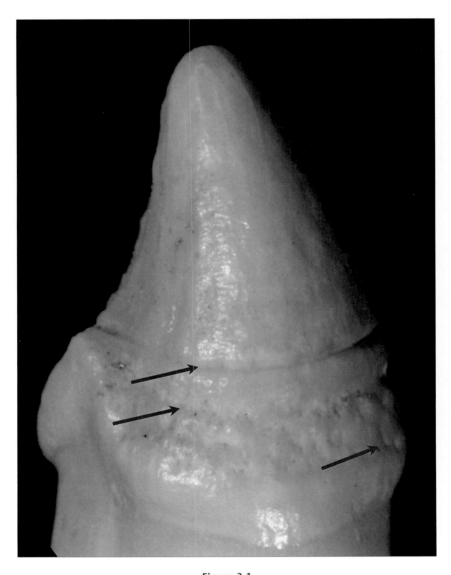

Figure 3.1
Developmental defects known as hypoplasias on the surface of a canine crown.
Note the circular ring (upper arrow) as well as the pitted areas (lower arrows) that
mostly encircle the tooth. Chimpanzee specimen courtesy of Christophe Boesch.

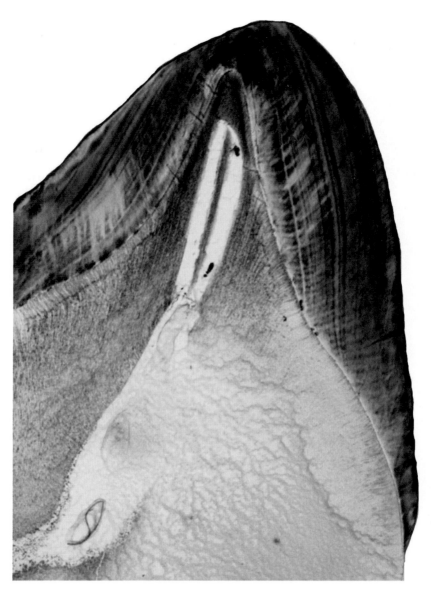

Figure 3.2
Early life stress in the first molar tooth of a human. The neonatal line (green arrow)
is followed by several dark, contrasted, accentuated lines (red arrows) formed
during the first year of life. Human tooth courtesy of Robin Feeney.

Table 3.1
Some Suggested Causes of Developmental Defects
(Hypoplasias and Accentuated Lines)

Type	Specific Cause
Nutritional	Malnutrition
	Vitamin A deficiency
	Vitamin D deficiency
Disease	Chickenpox
	Diabetes
	Diphtheria
	Measles
	Neonatal haemolytic anemia (immune blood disease)
	Parasite infection
	Pneumonia
	Rubella
	Scarlet fever
	Scurvy
	Smallpox
	Syphilis
	Whooping cough
Symptom	Allergies
	Bone inflammation
	Convulsions
	Diarrhea
	Fever
	Vomiting
Event	Antibiotic administration
	Birth
	Birth injury
	Electrical burns
	Enclosure (room) transfer
	Eye injury
	Fluorochrome (biomarker) administration
	Hospitalization (admittance or discharge)
	Inoculation (with pathogenic virus or bacteria)
	Ionizing radiation
	Parturition (giving birth)
	Trauma
	Weaning (cessation of breastfeeding)

seemed to match up well with the estimated ages of accentuated lines. These were formed within the same week as the unfortunate symptoms, and in several cases they happened on the exact day. The story of these twins' early lives is a striking example of the intimate recording system in teeth.

Studies of young primates raised in captivity reveal similar impacts on their tooth development.[5] For example, one unlucky gorilla showed a precise age match between accentuated lines and an eye injury and surgery. Subsequent hospitalizations and moves to new enclosures matched the ages of lines formed prior to this individual's accidental death. A few years ago, I conducted a study of developmental stress in a young monkey's molar, finding accentuated lines that corresponded to an enclosure transfer, a leg injury, and a tail amputation. Other lines marked bouts of dehydration and diarrhea that led to multiple hospitalizations. Another study of macaques reported evidence in their teeth of maternal separation, anesthetization, and handling for routine measurement. I find it surprising that simply moving these young primates to new rooms or enclosures created lines in teeth that mimic their illnesses. These experiences have measurable effects on youngsters' development, pointing to a degree of environmental sensitivity that their caregivers may be unaware of.

Intentional manipulations of other laboratory animals during the last century provide important insight into more severe defects on tooth surfaces.[6] At the time, there was considerable interest in understanding rickets, a disease that impairs skeletal formation in children. Rickets was common in northern cities, where exposure to sunlight is more limited. This was especially problematic during the height of the Industrial Revolution, when air pollution became extreme. Researchers used dogs as an experimental model, decreasing vitamin D to mimic rickets, then inadvertently discovering that this deficiency produced hypoplasias in their teeth. Similar nutritional and hormonal manipulations of mice led to dental defects, as did excessive fluoride supplementation and intentional parasitic infections in sheep. Another troubling example is the case of human children born with syphilis. This potentially fatal disease, which is caused by the movement of bacteria from an infected mother into the developing fetus, produces extreme hypoplasias and dental deformities (figure 3.3).[7]

Studies of these defects in living individuals help to guide interpretations of extinct species. One example I enjoy sharing with my students comes from a study of fossil ape teeth discovered in Turkey.[8] While surveying the

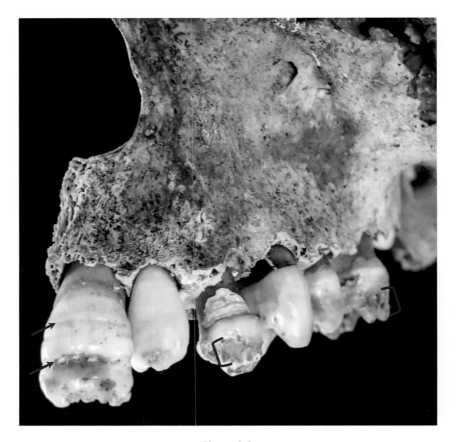

Figure 3.3
Hypoplastic defects in an 8- to 10-year-old human child believed to be born
with congenital syphilis. Notice the horizontal hypoplasias and the pitting on
the central incisor, deciduous canine, and first molar (arrows and brackets);
these regions were formed during the first few years of life. This individual was
buried in the "pauper section" of a cemetery with individuals who migrated to
Australia from the United Kingdom prior to the 1900s. Image credit: Stella Ioannou.

preserved dental remains, Jay Kelley initially noticed a similar pattern of
hypoplasias encircling the crowns of ten upper central incisors. Because
apes and humans have only two central incisors, he reasoned that these
teeth must have belonged to at least five different individuals. When Kelley
examined them under a microscope, he noticed that there were two hypoplasias on the same location on each of the incisor crowns—an important
clue to when they were formed. Kelley counted long-period growth lines

between the hypoplasias and found the same number of lines on all of the teeth, meaning the time between hypoplasias may have been equivalent. He combined these pieces of evidence to conclude that these apes were all at the same maturational stage, and that they experienced two stressful episodes simultaneously. From this, he inferred that these animals were part of a cohort of animals born at the same time. This means that their mothers would have reproduced seasonally—a behavior common to modern monkeys who live in temperate environments, but uncommon in living apes. Finally, because the incisors showed identical amounts of wear, he surmised that these individuals all died at the same time. Reconstructions of the site where the fossils were discovered support this interpretation. The geology indicates that the fossils accumulated rapidly, possibly due to a flood. Kelley's study demonstrates that incremental features don't only capture the "intimate history of the individual"[9]—they can also shed light on a whole group of apes that lived millions of years ago.

Biological anthropologists and archeologists have been especially interested in using these developmental defects to study breastfeeding practices in ancient hominins and modern humans.[10] As we'll discuss in chapter 6, knowing when mothers wean their babies helps us to understand how energy is used for growth and reproduction over the life course. From the weanlings' perspective, the end of breastfeeding is thought to be stressful since the flow of nutritional benefits and maternal antibodies ceases, along with the psychological comfort of nursing. Weaning too soon, without sufficient resources, can lead to malnutrition, and consuming nonmaternal foods carries a risk of infection from food-borne pathogens. Although weaning is frequently inferred as a cause of hypoplasias, evidence from humans is not conclusive. For example, the skeletal remains of 27 children of enslaved African Americans showed hypoplasias formed between about 1.5–4.5 years of age, yet historic reports indicate they would have weaned between 9–12 months. We're not sure what might have caused these dental disruptions over the following few years, although one can imagine that they faced innumerable hardships. I've also studied nonhuman primates with documented histories and have not found any proof that hypoplasias form during their weaning process. However, there is hope for determining weaning ages in the past, and in chapter 7 we will learn how teeth make this possible.

Part of the trouble in making sense of hypoplasias or accentuated lines is the fact that proving causation is more difficult than showing that two things are found together more often than would be expected by chance. This may seem like a small point, but there are numerous examples where an apparent link or association between observations has led to a mistaken belief that one thing caused another. For example, we know that some human societies had a weekly "feast day" until recently, and that human teeth show a long-period rhythm that often repeats on a weekly basis. This similarity in timing led some to speculate that feasting caused long-period rhythms. Today we know that having a large meal does not produce a line in dental hard tissues or bones. Humans who don't over-indulge on a weekly basis still show near-weekly rhythms, and those from populations that feasted on a weekly basis often show rhythms that are shorter or longer than a week. In a similar way, although teeth often show hypoplasias between 1–4 years of age—when mothers in many societies cease breastfeeding children—this association does not prove a direct causal relationship.

Anthropologists have also investigated whether malnutrition causes developmental defects.[11] Here we find more evidence that deserves further study. For example, rural Mexican and Guatemalan children given food supplements showed fewer hypoplasias than children who did not participate in the nutrition study. Scientists have also documented increases in hypoplasias during major cultural transitions, and point to potential deficiencies in the nutritional status or health of these populations.[12] The invention of agriculture beginning 10,000–15,000 years ago is the most extreme shift in nutrition and health over the history of our species. Prior to this time, our ancestors obtained food through hunting and collecting wild foods. During the initial "Agricultural Revolution," people began to cultivate cereals, rice, and other plants. They settled into permanent dwellings to tend crops and led more sedentary lifestyles. These early agriculturalists decreased the breadth of their diet, ate more carbohydrates, and lived in larger communities where diseases could spread more easily. Perhaps unsurprisingly, their teeth show more hypoplasias than hunter-gatherers.

Other cultural transitions suspected to show up in teeth include periods of social stratification or contact between different populations. Biological anthropologist Judith Littleton documented hypoplasias on the teeth of Australian Aboriginal children born between 1890 and 1960.[13] Their dental defects increased as social contact with Europeans became more frequent. This trend was particularly noticeable during World War II, when certain Aboriginal communities were relocated to government settlements and forced to cease hunting and gathering. Although rations were provided, limited nutritional quality and cramped living conditions led to poor health, and hypoplasias appeared at earlier ages and more frequently than in pre-contact populations. It would be worth exploring whether other indigenous communities show similar records in their teeth when giving up their traditional livelihoods.

Importantly, though, these examples do not establish that poor nutrition causes developmental defects. Malnourished children have weakened immune systems and are more prone to infections.[14] Given the apparent sensitivity of developing teeth, additional research is needed to untangle the potential interplay between subsistence methods, dietary quality, and immune function. This is an essential step that must precede deciphering the childhood hypoplasias in fossil hominins or prehistoric humans. Until then, we would be wise to refrain from settling on any particular interpretation without additional evidence. While this is less satisfying than arriving at a firm conclusion, documenting the general well-being of individuals thousands or millions of years after they died is a pretty respectable first step.

Those Pesky Wisdom Teeth

The final rite of passage for our dentition is the formation and eruption of our third molars. Recall that these appear during the transition to adulthood around 18 years of age—inspiring the moniker wisdom teeth. For centuries, we have suffered from two third molar disorders: failure to form, known as *agenesis*, and failure to erupt into the mouth, known as *impaction*.[15] Agenesis is easier to diagnose than impaction—it's simply the complete lack of a tooth. If impaction is suspected, we must rule out the possibilities that a tooth may be not be in place because it hasn't finished forming inside the jawbone or because it never formed in the first place. When teeth don't arrive on schedule, assessment of X-ray images is essential to figuring out what's going on (figure 3.4). These conditions affect one out of every four people living today, although they distress certain groups more often than others. It's very common for Asian, Middle Eastern and Native American people to be missing third molars, while Africans and Australian Aboriginals show lower incidences of agenesis. Similarly, impaction affects Middle Eastern and Asian individuals more often than African populations. European people fall in the middle for rates of agenesis and impaction, although

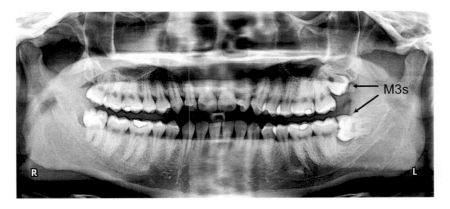

Figure 3.4
Panoramic X-ray showing impacted third molar teeth (M3s) in a living
human. Bright regions in other teeth are fillings. Image credit: Ka-ho Chu.
Modified from https://en.wikipedia.org/wiki/Wisdom_tooth#/media
/File:Impacted_wisdom_teeth.jpg.

there is considerable variation in individuals from different backgrounds
within these regions.

Missing and impacted wisdom teeth are thought to be recent developmental disorders.[16] Aside from a few potential cases in the distant past, these
conditions rarely troubled any hominin species other than our own. So why
are they so common in humans living today? Kate Carter, my former graduate student at Harvard University, set out to answer this question by testing
several long-standing hypotheses. One reason commonly given for agenesis
and impacted wisdom teeth is the belief that our jaws have shrunk faster
over evolutionary time than our teeth have, leading to arrested or stunted
third molar development. This idea invokes our recent soft-textured modern
diet—which requires less force while chewing than harder ancestral diets—as
a primary cause of jaw shrinkage. Recall that the Agricultural and Industrial
Revolutions, two of the most profound periods of human dietary change,
occurred fairly recently. Another hypothesis points to large-scale changes
in our diet that have lessened tooth wear, leading to more variable dental
development. A final idea is that natural selection led to the loss of third
molars due to the health risks of impaction. This theory holds that before
modern medical intervention, adults who retained their wisdom teeth experienced higher mortality from infections than those who never formed them.

Kate wanted to tease apart these hypotheses, so she lugged a portable dental X-ray unit to museums in Europe, Japan, and Washington, DC to study skeletons from people who lived during the development of agriculture and industry. She spent several months gathering information on tooth and jaw sizes, tooth wear, presence or absence of third molars, and impaction. Kate quickly confirmed that both agenesis and impaction were less common in her six archeological populations when compared to humans living today. Figuring out why has proven to be a bigger challenge for us both.

As one would expect, third molars get stuck more often in small-jawed individuals than in large-jawed individuals. In other words, wisdom teeth don't erupt properly when there isn't room for them, and this has seriously plagued humanity over the last few hundred years. Kate found a fivefold increase in third molar impaction in living humans compared to early agri-culturalists. Others have estimated that it has become 10 times more likely since the Industrial Revolution began. And impaction affects women, who tend to have smaller jaws, more often than men. Impaction of any tooth type also occurs with some regularity in captive and domesticated animals, which typically eat a softer diet than their wild counterparts.[17] The logic here is that bones respond to mechanical pressure, so consuming a harder diet during childhood leads to larger, stronger jaws. The uptick in third molar impaction may have occurred when humans abandoned hunting and gathering in favor of agriculture, beginning in the Middle East and Asia many thousands of years ago. It then reached even greater levels in those populations whose recent ancestors embraced industrially produced foods. Thus wisdom tooth impaction seems to be a true disease of civilization, or a mismatch between our modern culture and biology. This phenomenon is called regressive evolution or dysevolution—and third molars aren't the only part of our body to fall out of step with evolutionary design.[18]

Those of us who've experienced the unpleasant reality of wisdom tooth impaction firsthand understand why some dentists and oral surgeons often recommend removing them altogether. I had all four of mine extracted at age 20 after developing a rapid and painful infection in my lower jaw, and I was relieved to be spared any further problems. Infections of the gum or jaw can spread to the head, neck, or bloodstream, and can be fatal when left untreated.[19] This disproportionately impacts people who cannot afford dental care, as journalist Mary Otto documented in *Teeth*—her gripping book on the current oral health crisis in America. Even surgical removal of the third molars once entailed significant risk. A 1936 review of 622 cases of impacted wisdom teeth reported 42 deaths due to complications after surgery. Modern clinical practices are considerably safer, which is good news, since impacted wisdom teeth are likely to torment us for years to come.

Making sense of third molar agenesis is more difficult than figuring out why impaction has become so common. Kate was surprised to learn that larger-jawed individuals who lived during the agricultural transition were more likely to be missing their third molars than smaller-jawed individuals— the opposite pattern of what she found for tooth impaction and jaw size. This suggests that it wasn't simply the lack of space that was inhibiting tooth formation. Yet during the industrial transition, smaller-jawed individuals were more likely to be missing their third molars. Kate concluded that this second result is consistent with the theory that natural selection might have been acting to favor individuals who were missing third molars, particularly as people with small jaws who retained them were at greater risk of impaction, infection, and possibly death. Future studies of the overall health of individuals who lived during the development of agriculture and industrialization might help to confirm these conclusions. For example, gum and tooth infections often leave a characteristic signature in the bones around teeth—clues that we'll explore in the following section.

Cavities, Gum Disease, and Tooth Loss: Additional Diseases of Civilization?

After our teeth finish their slow passage into the mouth, native microorganisms can create an unwelcoming environment.[20] Bacterial colonies attracted by moisture rush to coat the pristine crowns, producing a "furry" feeling surface. You may have had the experience of running a fingernail or toothpick over your front teeth first thing in the morning, and noticed the sticky plaque that formed overnight. This "biofilm" is made of bacteria, proteins from our saliva, and microscopic food particles. Left alone over time, the innermost layers of plaque will begin to mineralize and form tartar, or *calculus*, which requires professional care for safe removal. While excess plaque and calculus threaten our oral health, in many ways they are a boon for anthropologists. I'll share in chapter 7 how calculus research is deepening our understanding of ancient diets. Calculus also traps bacteria, allowing a glimpse of certain diseases prior to recorded history.[21]

Dental plaque is the main culprit in cavity formation. The bacteria in plaque feed off saliva, fluid from our gums, and sugars we've ingested— producing acids that contribute to the destruction of our dental hard tissues. Strains of bacteria such as *Streptococcus mutans* (*S. mutans*) are particularly effi- cient at breaking down certain carbohydrates and forming lactic acid, lower- ing the pH of plaque. A cavity begins when this increasing acidity causes the loss of nearby minerals, forming a weakened area called a lesion. As the lesion undergoes further exposure to acid, the underlying enamel and dentine begin to break down, eventually forming a hollowed-out region, or cavity (figure 3.5). When left untreated, entire crowns may be destroyed, and roots may become so degraded that their bony attachments loosen, leading to tooth loss.

Cavities are often described as a disease of our modern lifestyle, although whether they are of modern origin or the remnant of an ancient scourge is still up for debate.[22] Scholars who favor a recent origin often point to the work of Christy Turner, who meticulously reviewed the prevalence of cavi- ties in archeological populations that practiced different subsistence styles.[23] Turner found that the frequency of cavities was very low in hunter-gatherers (2% of 47,672 teeth examined), higher in a mixed-subsistence group (4% of 58,137 teeth), and even more elevated in agricultural groups (9% of 504,095 teeth).[24] Complementary evidence comes from a genetic study of modern strains of *Streptococcus*, which estimated that the cavity-causing *S. mutans* became common sometime between 3,000–14,000 years ago. This period includes the origins of agriculture, a dietary transition that created favorable environments for *S. mutans* in the mouths of early farmers.

A study of Europeans who lived during the last few thousand years reached a slightly different conclusion. Dental calculus scraping of 12 indi- viduals from 4,000–7,600 years ago did not contain *S. mutans*, and it only turned up in a few of the more recent individuals. Given that *S. mutans* is very common in humans living today, the authors concluded that it became widespread after the Industrial Revolution. Although these stud- ies implicate different cultural transitions, they all link the prevalence of *S. mutans* to dietary shifts that may enhance its role in cavity formation.

Those on the other side of the debate note that this condition predates both agricultural and industrial practices. Cavities occur with some frequency in ancient hunter-gatherers, fossil hominins, and chimpanzees—our fruit- loving closest living relatives (figure 3.5).[25] A striking example comes from the first fossil hominin discovered in Africa, a member of the genus *Homo*

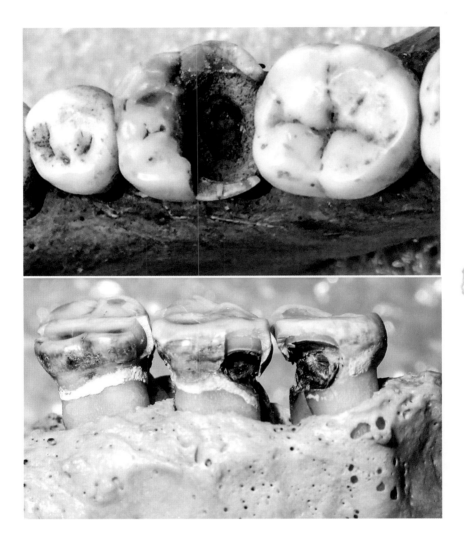

Figure 3.5
Cavities of the crown (above) and root (below) in molar teeth. The upper image
shows a modern human, and the lower image is a wild chimpanzee. The receded
bone beneath the chimpanzee crowns shows indications of periodontal disease.
Human jaw courtesy of the Phoebe A. Hearst Museum of Anthropology and the
Regents of the University of California (Catalogue No. 12-6971B.2). Chimpanzee
specimen courtesy of Christophe Boesch.

that lived 125,000–300,000 years ago. This fossil, known as the Broken Hill skull, shows cavities in 10 of its remaining 11 upper teeth. More recent hunter-gatherers were plagued by them as well. In chapter 9, we'll consider the case of a 14,000-year-old European hunter-gatherer who is the oldest-known person with a drilled cavity. Another study of ancient North Africans documented an exceptionally high incidence of cavities—51% of the teeth of 52 adults showed them, similar to the level of industrialized populations.[26] These 13,700- to 15,000-year-old humans likely relied on wild starchy foods including acorns and pine nuts, which may have contributed to their poor dental health. It would be helpful to check whether *S. mutans* is present in the calculus of these ancient hunter-gatherers, which might clarify the hypotheses discussed above.

Regardless of when cavities originated, rates in modern children and adults have skyrocketed, particularly with the advent of commercial sugar production during the Industrial Revolution.[27] Britain was one of the first major importers of sugar, and once the "sugar tax" was reduced in 1874, it became more affordable for the general public. The remains of British individuals who lived over the last few hundred years show fivefold increases of cavities in children and threefold increases in adults compared to their predecessors.

Another striking example comes from a comparison of two contemporary Mayan communities in Mexico.[28] One village relies on a traditional agricultural economy dominated by the cultivation of corn (maize), while the other village has access to global food markets, which include processed foods and artificially sweetened drinks. When individuals in each village were compared, those in the village with processed foods had more cavities in 9 of 12 comparisons of the same sex and age groups. To be clear, cavities affected more than half the adults in both villages, but the authors attributed the nearly ubiquitous incidence in the global-market village to the popularity of refined sugars and sodas. Sadly, this phenomenon is playing out across the world as many traditional rural communities join the global economy.

These comparative studies, as well as numerous experimental ones, have demonstrated that particular diets cause cavities, yet other factors also play a role.[29] For example, women tend to develop cavities more often than men. One folkloric adage "gain a child, lose a tooth" points to the belief that childbearing exerts a special toll on the oral health of women. Differences in traditional food gathering and preparation roles may lead to more snacking by women. Furthermore, women have a higher amount of estrogen than men, a hormone believed to affect saliva production. Saliva is a key part of the body's defense against bacteria, helping to keep the teeth free from sugars, buffered from acids, and bathed in antimicrobial compounds and minerals that help to harden teeth. Men may experience more protection from their saliva, and this difference becomes even more pronounced for pregnant women, since bacterial levels are often elevated during pregnancy. Changes in diet and oral health over each trimester may be a perfect storm for *S. mutans* to induce or accelerate cavity formation—particularly when woman bear large numbers of children. In this case, traditional folklore appears to have been validated by modern science.

The good news is that cavities rarely lead to fatalities for people with access to modern medical care. At their worst, they create painful infections of the pulp and supporting tissues, as well as tooth loss.[30] Yet cavities aren't the only reason we lose teeth. Prolonged gum inflammation—driven by plaque and calculus buildup, sugary diets, stress, smoking, genetics, and certain bacteria—is a major reason teeth fall out.[31] This begins when invading bacteria trigger an immune response, causing inflammation that leads to the loss of the surrounding periodontal tissues over time, particularly when these other factors are present. Diagnoses of periodontal disease in adults living today vary greatly depending on how it is defined. One estimate suggests that the early inflammatory stage, gingivitis, affects 90% of adults worldwide.[32] Like many Americans over the age of 30, I've been warned that I need to be vigilant about my oral care, as my gums have begun to show the telltale inflammation and recession that could lead to periodontal disease—and ultimately, to tooth loss.

Of greater concern is the fact that periodontal disease is linked to life-threatening diseases, including diabetes, heart disease, and cancer. While these medical conditions don't typically leave traces in fossils, the presence of diseased bone is an important indicator of poor health. Biological anthropologists have also pondered whether periodontal disease has ancient roots.[33] Possible cases from the hominin fossil record date back to 1.8 million years ago in Eurasia, and include several individuals recovered in Spain from over a span of one million years.[34] In these instances, the bone around their tooth roots shows resorption, or loss, thought to be a consequence of prolonged gum inflammation. Studies of recent humans suggest that the bacterial strains associated with periodontal disease have been relatively stable since the introduction of agriculture, yet dental remains recovered from archeological digs show less periodontal disease prior to the Industrial Revolution than is found in modern populations. Here again it seems that there may be more than a single factor contributing to dental disease.

Additional research is needed to clarify how current levels of periodontal disease are related to bacterial prevalence and dietary changes. An international research team was invited to supervise a creative experiment that bears upon the issue of oral health in the past.[35] Ten adventurous Swiss people were recruited to undertake four weeks of primitive "Stone Age" living in an archeologically informed environment, including simple huts, clothing, and mostly wild foods. Their health was evaluated at the beginning of the study and again at the end. The intrepid group was given a basic supply of whole grains, as well as salt, herbs, honey, milk, and fresh raw meat. The participants supplemented their provisioned food with locally foraged items, and they had to prepare meals themselves without modern implements. Barring access to modern toothbrushes, dental floss, or toothpicks, some attempted to clean their teeth with twigs. Interestingly, the group members' gums were not more inflamed than when they began four weeks earlier, although their plaque levels were elevated. While the design of the study would leave most academics skeptical—particularly as it formed the basis of a Swiss reality television show—the elimination of refined sugars may have counterbalanced their primitive oral-care routines. Over four weeks the participants' oral bacteria communities changed, and certain strains associated with gingivitis became less prevalent while other microorganisms flourished. The team concluded that a lack of dental care does not necessarily lead to gingival inflammation, particularly when processed foods are avoided. I'm not so sure that my favorite dental hygienist would agree!

A final aspect of oral health that is evident in ancient hominins and prehistoric humans is tooth loss. Prior to the advent of agriculture, tooth loss due to heavy wear was quite common.[36] Once agriculture intensified, tooth wear lessened but cavities increased—becoming the main cause of tooth loss. When wear or decay reaches the pulp space inside the crown, the soft tissues become susceptible to bacterial invasion and infection. As the pulp becomes inflamed, its blood vessels, nerves, and dentine-forming cells die. The infection may then spread through the root canal and out into the surrounding ligament, loosening the connection between tooth roots and their bony sockets. Infections may also destroy the nearby bone, creating sizable holes when left untreated (figure 3.6).

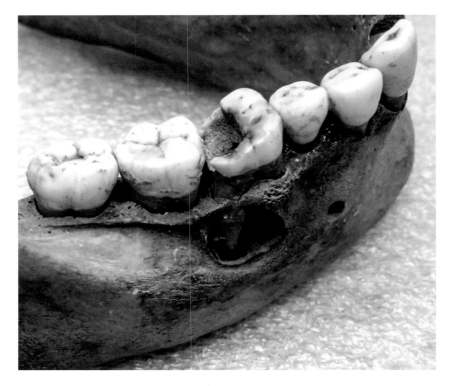

Figure 3.6
Bone infection (abscess) likely due to a cavity (see figure 3.5) in a modern
human lower jaw. The resorption of bone beneath the crown is also indicative of
periodontal disease or gum inflammation due to the infection. Human jaw courtesy
of the Phoebe A. Hearst Museum of Anthropology and the Regents of the University
of California (catalogue no. 12-6971B.2).

Once teeth are lost, the body resorbs the bone that surrounded the roots,
and jaws become shorter from top to bottom (figure 3.7). A jaw that has
lost its teeth is known as *edentulous*, meaning "out of teeth." Over time, it
becomes impossible to determine if teeth were lost because of heavy wear,
cavities and pulp infection, periodontal disease, or intentional removal—
which we will learn about in chapter 9. Several ancient hominins are
known to have partially edentulous jaws, including a 1.8-million-year-old
individual who only had a single tooth left at death.[37] A classic example
is "The Old Man" from La Chapelle-aux-Saints, an arthritic French Neanderthal retaining only a few front teeth. Their discoveries have inspired
lively discussion about whether ancient hominins took care of their elderly

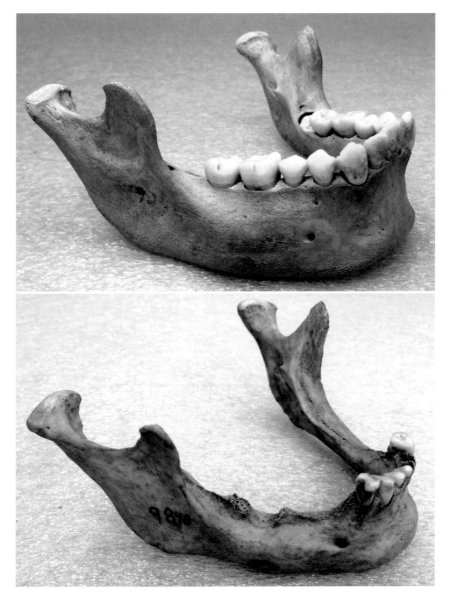

Figure 3.7
Comparison of a healthy human adolescent jawbone (above) with a partially
edentulous jawbone (below) from an aged adult. Human jaws courtesy of the
Phoebe A. Hearst Museum of Anthropology and the Regents of the University
of California (catalogue nos. 12-9042(0), 12-9840(0)).

members, since toothless individuals would have had difficulty eating wild foods that hadn't been softened or cooked. While our oral health has been negatively impacted by the recent adoption of a carbohydrate-rich diet, these hunter-gatherers didn't have it easy either! Heavy wear, infections, toothaches, and lost teeth appear to have been a fact of life for hominins, particularly our own species—as we've been surviving to reach "old age" more often those that preceded us.[38]

A recent economic study of lost productivity ranked tooth loss as the most costly dental problem across the globe.[39] Global incidences in 2010 were estimated to have led to the loss of $63 billion U.S. dollars, edging out periodontal disease at $54 billion and adult cavities at $25 billion. Clinicians and dental researchers have long endeavored to create affordable replacements. Although artificial teeth—including modern bridges and dentures—have improved the quality of life for millions of people, tooth replacement has had a rather gruesome history. In the eighteenth century it was common for destitute people to sell their healthy teeth to dentists, who would yank them out and implant them in the mouths of wealthy individuals.[40] Victor Hugo's novel *Les Misérables* dramatized this unsavory practice, which wasn't particularly successful, since teeth require intact blood vessels, nerves, and ligaments to function properly. Moreover, transplanted teeth often carried syphilis, leading to dangerous infections in their new owners. In the concluding chapter we'll learn about cutting-edge efforts to "bioengineer" brand new hygienic teeth.

Use It or Lose It: Evolutionary and Behavioral Perspectives on Orthodontics

Over my career, I've had the chance to examine the dentitions of hundreds of great apes and fossil hominins. Despite the occasional cavity or lost tooth, these groups tend to have neatly fitting teeth. Yet a quick look in my mirror or at a smiling crowd watching a comedy show should give hope to aspiring orthodontists. Nearly two-thirds of all Americans show *malocclusion*, or misaligned teeth, yet another disease of modern civilization.[41] Many of us have fretted through our preteen years with awkward metal braces, while the younger among us perhaps decided to straighten teeth with removable Invisalign trays. It might surprise you to know that malocclusion treatments over the last century have been heavily influenced by evolutionary theories about dental crowding, and recently these have been called into question.

In the 1920s, the up-and-coming orthodontist Percy Raymond Begg conducted an influential study of Australian Aboriginal hunter-gatherers, who were known to have large and heavily worn teeth.[42] Begg noticed that malocclusion was less common in their teeth than in other contemporary human populations. He surmised that excessive tooth wear helped to reduce dental crowding. This is because teeth do not just wear down on the chewing surfaces. Wear also occurs between adjacent teeth as they rub against one another, which eventually shortens the length of the whole tooth row. This creates room for teeth to drift slightly forward within the jaw as they continue their slow process of eruption. Begg concluded that humans consuming soft diets suffered from having overly large chewing surfaces, crowding jaws during adulthood, since their diets didn't create sufficient tooth wear. He regarded the "textbook model" of unworn human teeth in perfect occlusion as unnatural—a bold and controversial position to take as an aspiring clinician. This conviction led him to begin removing premolar and molar teeth from his patients in order to create space for other teeth to properly fit together, which remains a common orthodontic practice today.

An alternative evolutionary hypothesis for malocclusion has come from observations of face, jaw, and tooth size during the Agricultural Revolution. Comparisons of ancient Egyptians show that the facial skeleton decreased in size and shifted back toward the spinal column as their reliance on agriculture increased.[43] As we've discussed, the diets of many agriculturists were less diverse than hunter-gatherers, with a heavy consumption of cereals and starchy plants that were processed into soft digestible forms. Scientists extrapolated from the patterns in Egyptians and our fossil ancestors that reductions in chewing stresses led to the development of smaller faces. Thus, the problem for modern humans is not that we don't wear our teeth down enough— rather, it's that we don't grow our jaws large enough to hold our teeth.

This idea is further strengthened by comparisons of teeth and diets across generations of recent humans.[44] When older individuals who grew up eating a harder, more traditional diet are compared with younger generations who eat softer processed diets, the younger folks show greater malocclusion. Similar trends are found when comparing rural Indian children on a traditional diet with those that grew up in an urban environment with more processed foods. Moreover, animals fed soft diets show reduced facial sizes and crowded teeth when compared to their wild counterparts.[45] These comparisons support the idea that misaligned teeth are caused by the lack of sufficient chewing forces during development. They are also consistent with links between jaw size and tooth impaction, as recent small-jawed individuals are more likely to have impacted third molars. The notion that the jaws of modern children aren't growing large enough to hold their teeth is starting to gain traction in clinical and evolutionary communities.[46] Some kids are being offered jaw-strengthening exercises as orthodontic experiments, which wasn't an option for my generation of braces-baring teenagers. I'll be fascinated to learn if this new interpretation of the past can help us to overcome a disease of modern civilization—sparing others like me the physical and social discomfort of having their teeth corrected.

Armed with a foundation in how human teeth develop, change with age, and are subject to disease, we're now ready to explore their origin and evolution. We'll start roughly 500 million years ago, situating humans first as vertebrates, then mammals, and most narrowly as primates—evolutionary radiations that show incredible diversity in tooth form and function. In the following chapters, you'll learn about key adaptations that led to the evolution of mysterious hominins, who populated the earth over the past seven million years. We'll also see how teeth provide important markers of our remarkable life cycle and complex cultures—distinguishing us from our ancestors and enabling our survival as the sole hominin left on the planet.

II   Evolution

I am more interested in transformational aspects of evolution than in bifurcation aspects.

—Percy M. Butler, PhD, 1996 (quote from published remarks upon receiving the Romer-Simpson Medal, *Journal of Vertebrate Paleontology* 17 (1997): 248–250.)

4 The Fish to Primate Transformation

For insight into human evolutionary history—especially from the perspective of our dental development and anatomy—we need to step back in time to the origin of vertebrates. This is the point where scientists have detected the first adaptations that eventually gave rise to our pearly whites. The evolution of what I've casually termed our oral Swiss Army knife is a bit like humans' oversized brain—the ultimate driver of our behavior. Simply put, our brain has a reflexive reptilian core, a nurturing mammalian middle, and a prosocial primate coating. Our teeth evolved from a simple peg-like motif that first allowed fish and reptiles to prey upon simpler organisms, and one another, while populating the seas and landmasses. This was followed by changes in the development and design of mammalian dentitions, which were further refined into an omnivorous ape pattern. We'll trace these major evolutionary steps— as well as a few diversions into the realm of dental oddities—leading to the evolution of our slow-growing pearly whites. Although of ancient origin, teeth continue to excite biological anthropologists who discover new species—ranging from tiny primates in the modern deserts of Wyoming to massive apes in mountainous limestone caves that dot vast Chinese fields.

The First Teeth: Eat or Be Eaten

The story of the earliest teeth takes place in a murky undersea world populated by some of the first animals with internal skeletons, known as *vertebrates*. Today's vertebrates encompass creatures we instinctively think of as "animals," including most of our beloved pets. Yet the marine world was alive with other animals prior to the evolution of vertebrates. The ancestors of modern snails, jellyfish, and insects had been around for more than 100 million years before our skeletonized forebears appeared.

Vertebrates are named for locking spinal bones—or vertebrae—that serve as a rigid support for muscles and other bones that protect the brain, heart, and digestive system. Refinements of the earliest skeletons also provided attachment sites for muscles that help streamline swimming in sharks and fish, and limbs that allowed four-footed animals to crawl onto land. An additional advantage was the skeletal and muscular support provided for a set of jaws—a major breakthrough for crushing food in an aquatic world that had hitherto relied on more passive forms of feeding.

You might think that with jaws came teeth, but this isn't necessarily so—and it depends on how you define teeth.[1] We've already encountered jawless snails that have simple pointed structures near their mouths that allow them to scrape food off rocks. Insects such as spiders have evolved fangs and other "teeth" that allow them to puncture their prey. Yet they are made according to a different mechanism than the one we encountered in chapter 1, and they have a separate evolutionary origin. Snails, spiders, and other invertebrates use particular elemental combinations to strengthen their feeding structures, which differ from the hard minerals that work so well in the teeth of primates and other vertebrates. Invertebrates often use calcium carbonate or the organic compounds keratin or chitin to harden their teeth. Chitin not only stiffens their teeth, it also enables the formation of elaborate "skeletons" to cover soft bodies. And as we'll see below, invertebrates provide an energy-rich snack for many animals, fueling excursions into new lands.

Outer skeletons were in vogue not just among invertebrates; some of the earliest jawless fish were covered in hard bony plates, or "armor skin," leading to the name ostracoderms. Hard reinforcements around their heads are thought to be one possible source of the first true teeth.[2] This is because small hard pegs dot the surface of this armor, a defensive adaptation that many scholars view as precursors to teeth inside the mouth. The inner structure of these "skin teeth" is remarkably similar to modern oral teeth—a hard dentine-like coating covers a softer pulp core, which is attached to a bony base. Tiny internal lines reveal their pattern of growth, much like the lines in our own teeth. Some fish species even have pegs that are covered by an enamel-like substance. Modern sharks and other living aquatic animals also have similar tooth-like structures on their scales.

One influential theory of tooth evolution suggests that the external skin teeth migrated inward to function as biting teeth in later-evolving jawed fish. For example, a 424-million-year-old fossil fish discovered in Sweden supports this theory. Scales on the outside of *Andreolepis'* jaw have enamel-covered bumps, and they lie next to rows of dentine-covered oral teeth.[3] In between the external scales and the internal pointed teeth is a transition zone complete with scale-like cones. These cones show chipping and cracking, which may indicate that they were used in biting, although they were not shed and replaced like the internal oral teeth.

You'll remember from chapter 1 that a special epithelial layer inside the mouth is an essential part of modern tooth formation. This layer is distinguished from other epithelial cells by the expression of key genes that send tooth-initiating signals to the underlying mesenchyme cells. An alternative theory about the evolution of teeth is that external skin teeth and internal oral teeth had separate evolutionary origins. In this view, the recipe for making a tooth is thought to be more malleable, leading to the independent genesis of similar tooth-like structures in multiple groups of early vertebrates. Part of the difficulty in choosing between competing hypotheses lies in the fact that the heads of these vertebrates vary in their appearance and development.[4] In addition, evidence during the critical period from 400–500 million years ago is woefully incomplete.

A classic example of how the fossil record can be misleading is the case of an extinct eel-like vertebrate given the moniker conodont, or "cone tooth."[5] Conodont teeth are the oldest known in the fossil record, and when they were originally found, they were thought to be complete tiny creatures (figure 4.1). Eventually scientists found fossilized impressions of long, soft bodies with similar teeth, and it became clear that the teeth were part of a larger organism resembling modern lampreys and hagfish. They lined the throats of these jawless animals, processing food as it passed into their digestive system. This feeding method is still used by living skates and rays, which crack shells between hard flat teeth in their throats. Paleontologist Philip Donoghue—a conodont expert—believes that these throat teeth did not lead to the evolution of teeth in jawed vertebrates, as they differ in their structure, function, and development.[6] If Donoghue is correct, it means that vertebrates have evolved biting teeth at least twice: once in jawless conodonts, and again in fish with jaws. In the latter case, having both teeth and jaws gave these vertebrates a serious advantage over ostracoderms and conodonts, since they could actively grasp and puncture their prey. These adaptations led to an evolutionary arms race, producing an

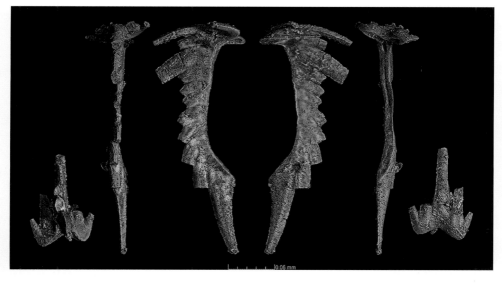

Figure 4.1
Conodont teeth found in fossil deposits in China. Image from http://paleo.esrf.eu
(Novispathodus_sp_TQ84C30_06_plate1.jpg). Originally published in Nicolas
Goudemanda, Michael J. Orchard, Séverine Urdy, Hugo Bucher, and
Paul Tafforeau, "Synchrotron-Aided Reconstruction of the Conodont Feeding
Apparatus and Implications for the Mouth of the First Vertebrates," *Proceedings
of the National Academy of Sciences USA* 108 (2011): 8720–8724.

incredibly diverse assortment of extinct fish, as well as ferocious modern
predators such as barracudas, piranhas, and sharks.

Modern fish teeth seem rather basic when compared to the varied shapes
of the teeth of humans and other mammals. The standard fish pattern is
one of uniform-looking conical pegs or barbs with a single root, which
may increase in size as they repeat along the tooth row. As with many
tales of dental evolution, exceptions abound—including the human-like
front teeth of some freshwater and saltwater fish, or the blade-like serrated
cusps of shark teeth.[7] Some marine vertebrates grow fanged teeth on their
tongues, or have multiple rows or plates of teeth inside their mouths. After
all, eating is one of the most important acts animals perform, and any small
change that makes it easier or faster is a real advantage in a competitive
world. Exploring the teeth of fossil and modern vertebrates has sparked my
curiosity about the almost unimaginable range of dentitions that arose and
spread through the process of natural selection.

Another important adaptation of fish is their continuous formation of replacement teeth.[8] Many fish grow slowly throughout their lifetimes, so it makes sense that they keep adding teeth as their jaws get bigger, as well as when they are damaged or dulled. In some cases, early-forming teeth are replaced by generations of more complex "adult teeth." Tooth replacement has been documented in some of the earliest fossil fish, which resorb the root where it attaches to the jawbone—detaching the old tooth while making space for newly minted recruits. Mammals undergo a similar process in order to shed their smaller and simpler baby teeth, although they typically have only one replacement set. In contrast, shark teeth naturally break off or fall out without tissue removal, with successive generations forming behind those that are in use (figure 4.2).

Having new teeth waiting in the wings, so to speak, is an ideal design to maintain a lethal dentition. Great white sharks—famously featured in the movie *Jaws*—can have several dozen 8-centimeter (3-inch) serrated teeth in their mouth at the ready![9] This process of continuous tooth replacement has captured the attention of scientists interested in bioengineering new teeth for human patients. Many of the developmental signals that control our own tooth formation can be traced to a common genetic recipe that stretches back more than 400 million years, and if we can learn how early vertebrates tweaked it, we might be able to nudge our own tooth formation to great advantage.

Land Ho!

The seas must have been a rough place to live during the evolution and diversification of jawed fish. This is particularly clear from records of decreasing oxygen levels and mass extinctions that occurred 360–400 million years ago.[10] Around this time, a group of intrepid fish found safer environments in shallow waters, paving the way for the evolution of four-limbed animals known as *tetrapods*. The gradual transition of animals from water to land involved a fundamental reorganization of their well-honed aquatic body plan and respiration system. While most fish obtain oxygen from water passing through gills, modern lungfish use gills and lungs to breathe—giving them the option to explore shallow pools and shorelines.

The ancestors of lungfish were important players in the transition to land, and some had large dentine-covered plates for crushing hard food

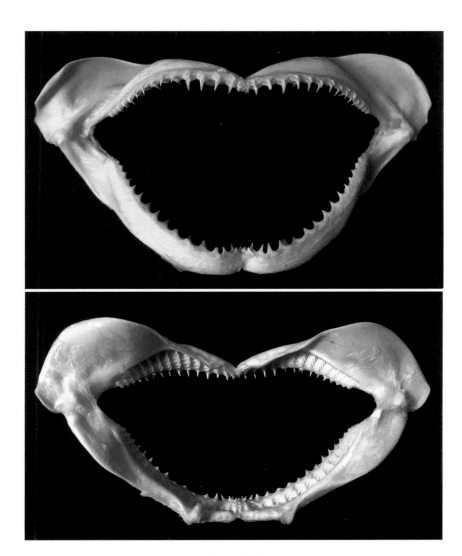

Figure 4.2
Shark jaws with simple teeth (top), with successive generations apparent from the
inner aspect (bottom).

found in coastal waters. Certain ancient fish also show an important restructuring of their front fins, which became pronounced 360–380 million years ago. A classic example of such a transition fossil is *Tiktaalik*, a remarkable animal with new adaptations for life outside water.[11] *Tiktaalik* had a long, flat head with a joint between the skull and the rest of the body—giving it a neck of sorts. It also had mobile joints in its front fins like the elbow and wrist of modern tetrapods. The discoverers suggested that these joints allowed *Tiktaalik* to do a kind of push-up, effectively lifting its head up and out of the water. However, its teeth, jaws, and scales show that it still had a lot in common with fish, leading one of its discovers to dub it a "fishapod."

Fossils that follow *Tiktaalik* in time show the gradual acquisition of additional tetrapod characteristics, including a strong rib cage for keeping the lungs from being crushed while on land, as well as limbs and a pelvis that strengthened the body and increased its mobility. These four-legged creatures began to explore the shores of ancient landmasses, but retained their easeful ways in the water, somewhat like the amphibious salamander. If you've ever watched a salamander walk, you might have noticed that they wiggle their body back and forth with their paired front and back limbs set diagonal, or opposite to one another. Preserved fossil tracks predating *Tiktaalik* show that an even more ancient four-footed creature crawled this way through the soft terrain of a shallow lagoon or tidal zone 395 million years ago.[12] By about 360 million years ago, the first bodies of true tetrapods appear in the fossil record. Their spiky peg-like teeth and simple hinged jaws allowed them to consume small fish and land-dwelling insects, although they probably weren't terribly graceful eaters.

Over the following 60 million years, newly forested lands teemed with insects and diverse four-footed vertebrates, many of which did not leave modern descendants. The two groups that survive today are distinguished by their method of reproduction. Modern amphibians—frogs, salamanders, and newts—require water for laying eggs, a remnant of their aquatic ancestry.[13] Like most water-dwelling vertebrates that preceded them, amphibians continuously replace their teeth. These are rather small and peg-like, although they may show some elaborations—particularly in adult salamanders. Amphibians specialize in eating insects and other invertebrates, often grasping and swallowing them whole, which doesn't require a particularly efficient dentition. In fact modern toads have completely lost their teeth, relying instead on their sticky tongues to capture prey and move it into their esophagus. The loss of teeth is a repeating theme in the story of vertebrate evolution: teeth shrink or disappear entirely in several types of animals.

The second modern tetrapod group includes reptiles, birds, and mammals, which carry their young internally or lay specialized eggs on land, and are known as *amniotes*.[14] Paleontologist Michael Benton describes the amniotic egg as a private pond—opening up a whole new world as amniotes were able to move away from water. But how do amniotes escape from their calcified or leathery cocoons once it's time to hatch? It turns out that many reptiles and birds grow a special "egg tooth" on the front of their heads that they use to free themselves, shedding this after they tap their way out during hatching. This seems like a fitting tribute to the ancient aquatic ostracoderms, with their hard bumpy heads.

Amniotes nurture their offspring, beginning an evolutionary trend of parenting that is a hallmark of mammals and especially primates. Fish and amphibians tend to produce many more eggs than amniotes, a strategy to ensure that at least a few offspring survive the perils of the aquatic world. In contrast, amniotes have fewer eggs, which they protect and care for to a greater degree. Biologists often distinguish between animals that "live fast and die young"—making many offspring as quickly as possible—from those that are slow-growing, long-lived, and focused on raising a few highly tended young. Humans are on the far end of this "helicopter parent" continuum, a reproductive strategy we'll discuss further in chapter 6.

The first amniote fossils appear around 310 million years ago as reptile-like creatures with long, flat skulls and sharp, peg-like teeth.[15] These slender, four-footed creatures may have initially subsisted on millipedes and other insects as they came to dominate the landscape, evolving diverse carnivorous and herbivorous forms over time. Everything was going well until an unprecedented mass extinction occurred about 60 million years after they originated, leading to a loss of more than half of all known amniotes. While the cause of this extinction is not entirely clear, volcanic eruptions contributed to increasing temperatures, dwindling atmospheric oxygen, and corrosive acid rain over many thousands of years. These significant environmental changes likely contributed to major disappearances of land plants, plankton, and marine invertebrates that are also apparent in the fossil record. If nothing else, amniotes experienced major disruptions in their food supply!

The following 50 million-year period of geological time is known as the Triassic, an era marked by the appearance of dinosaurs, modern lineages of reptiles, and small, warm-blooded mammals. Although the Triassic began with many fewer land-dwelling vertebrates than the previous period, favorable environmental conditions led to rapid evolutionary bursts of animals with novel adaptations for feeding, reproducing, and moving. Large reptilian carnivores with bladed teeth ruled the land, while crocodile-like creatures with large interlocking teeth hunted along the shores, and fierce-looking marine reptiles patrolled the seas. One exceptional reptile called *Placodus*, meaning "flat tooth," had projecting incisors to pry mollusks from rocks and thickly enameled large, flat back teeth with which to crush them (figure 4.3). *Placodus*

Figure 4.3
Skeleton of the marine reptile *Placodus*. Image courtesy of Wikipedia and Ghedoghedo (CC BY-SA 3.0; file: Placodus gigas 2.JPG). Fossil courtesy of the State Museum of Natural History (Stuttgart).

may be the Triassic's version of the sea otter, although perhaps not as cuddly as the playful oyster-eating mammal we know today!

Innovations in locomotion during the Triassic include the appearance of two-footed or bipedal stances and gaits, as well as a new four-footed posture with limbs tucked under the body—allowing for the efficient walking and running practiced by modern horses and cheetahs.[16] These changes paved the way for the evolution of dinosaurs, birds, and mammals. Dinosaurs are distinguished from other reptiles by their skeletal anatomy below the head, particularly in the specialized configuration of their pelvis and back legs. Their teeth and jaws largely retain the simple design and configuration of early reptiles, although some carnivorous groups had specialized serrations on their teeth, while others lost their teeth entirely. Recall that fish and fossil tetrapods tended to swallow their food whole rather than chewing it, which was also true of most dinosaurs.

One group of herbivorous dinosaurs evolved the ability to move their teeth against each other in a rotational scissor-like fashion, approximating a kind of chewing that typically distinguishes mammals from other vertebrates.[17] These herbivorous dinosaurs had special jaw joints and variable teeth with peg, dagger, and diamond shapes, yet another mammalian trait of growing multiple tooth types in the same row. This is an example of *parallel evolution*, which occurs when similar-looking features arise in multiple groups but are not inherited from a common ancestor. Another example is the loss of teeth in turtles, birds, and baleen whales. Each group adapted their dentitions in the same way, or in parallel, as they evolved. This "use it or lose it" process happened independently in each lineage, since the common ancestor of these groups possessed teeth. Alternative feeding mechanisms apparently provided advantages over their toothed predecessors.

Dinosaurs were a highly diverse group that dominated the landscape for almost 200 million years—capturing the imagination of generations of contemporary human kids. Researchers have identified growth lines in dinosaur bones and have measured the chemistry of their teeth to assess whether they were "warm-blooded"—employing internal temperature regulation like mammals and birds—or whether they were more similar to "cold-blooded" reptiles, with slower growth and metabolism.[18] These studies show that dinosaur growth rates and body temperatures were higher than in reptiles, although it is unclear whether large dinosaurs fall somewhere between cold-blooded and warm-blooded vertebrates, or whether all dinosaurs were metabolically equivalent to mammals. Additional support for the idea that they regulated their body temperature with an internal thermostat comes from growing evidence that some, if not all, dinosaurs possessed feathers, which provide a form of insulation. The development of temperature regulation may be a curse and a blessing. Warm-blooded creatures can stay active at all hours even in cold environments, but it takes an enormous amount of food to stoke that internal furnace on a long winter night—as those of us who are inclined to put on a few pounds during cold weather may be familiar.

Despite the fact that dinosaurs disappeared around 65 million years ago, their evolutionary descendants live on in the form of birds. A classic transition fossil is *Archaeopteryx*, one of several dinosaur-like bird groups from 150–160 million years ago that retain teeth, claws, and long tails.[19] Whole bodies of *Archaeopteryx* fossils have been found nearly flattened between plate-like rocks, which often preserve impressions of their feathers. The first true fossil birds had lightly built skulls with small teeth, which were serrated in some groups, blunt in others, and hook-like in a group that specialized in catching fish. Although modern birds are toothless, they still retain the potential to grow teeth, as revealed by experimental combinations of embryonic dental tissues.[20] Innovative studies have successfully paired dental epithelial and mesenchymal tissues from animals separated by more than a hundred million years of evolution. They prove that the formula for tooth formation hasn't changed all that much during amniote evolution. The process can even be restarted in a very common toothless bird.

Scientists have shown that a specific chicken strain known as *talpid²* begins developing teeth while it is incubating.[21] Unfortunately, this strain carries a genetic mutation that is lethal, meaning that the chicks do not survive long enough to be born. What would you expect chicken teeth to look like? The mutant chicks' teeth are conical and saber-shaped, growing like the teeth of baby crocodiles. This shouldn't be too surprising, since crocodilians are the closest living relative of birds, and they shared a common ancestor with simple conical teeth. This research provides rare insight into evolutionary processes, since the loss of teeth in birds appears to be due to a change in molecular signals rather than the loss of the genetic potential to initiate teeth.[22] Discoveries in this new field of evolutionary developmental biology help fill in gaps in our understanding of how teeth and other anatomical structures have come and gone over their 500-million-year history.

Vive La Révolution!

Humans and other mammals living today are united by a number of characteristics, including the possession of hair, mammary glands for nursing young, and longer juvenile periods than those of many other vertebrates. More subtle features are apparent in the design of the skull, including the ear, jaw, and teeth.[23] These traits did not appear all at once in a single fossil group; transitional reptiles that lived before the start of the Triassic period had some of these features, while others appeared piecemeal in extinct mammal-like creatures. The oldest definitive mammals appeared around 205–225 million years ago. These small, warm-blooded animals had large eyes, keen hearing, and pointy molar teeth, adaptations that helped them prey upon insects at night. We'll soon see how the earliest primates might have gotten a similar start millions of years later.

Important aspects of our own teeth can be traced to the fossil record of these enigmatic early mammals. One important modification is the development of varied tooth shapes that fit together in a tight interlocking manner— breaking down food quickly during chewing. Mammals have four tooth types from front to back: incisors, canines, premolars, and molars (figure 4.4). These

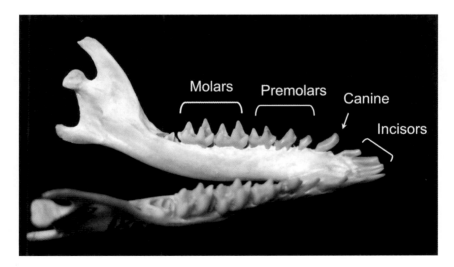

Figure 4.4
Mammalian dentition showing different tooth types, including multi-cusped molars. Tree shrew specimen courtesy of the Museum of Vertebrate Zoology, UC Berkeley.

shapes are determined by signaling molecules encoded in the dental mesenchyme. While other vertebrates occasionally evolved tooth shapes that differed throughout the tooth row, transitional reptiles sported a new kind of pattern.[24] Early mammals developed this further, producing multi-cusp molars that fit together in a precise way that both cuts and grinds food as the teeth meet. This works because the projecting cusp tips of upper molars fit in basins of the lower molars like a mortar and pestle, while adjacent cusps slide past one another, producing a shearing edge like a pair of scissors. The design is a landmark innovation for mammals, because it advantages them over other vertebrates who swallow their food largely intact. When food is cut into smaller pieces in the mouth, it can be digested faster and with greater energetic return. This then fuels milk production, brain functions, and the rapid metabolism required to maintain high body temperatures and active lifestyles. Humans have taken this dental milestone even further by crafting numerous tools for cutting and grinding food before we pop it in our mouths, which we'll discuss in the following chapter.

Part of the shift that aided the evolutionary diversification of mammals includes a period of infancy when newborn offspring rely on their mothers for food and immune support. These are supplied through energy- and antibody-dense milk produced in mammary glands. During the period of nursing, mammals erupt their milk teeth, or deciduous teeth. These teeth grow quickly, filling small jaws in time for youngsters to start chewing solid foods.[25] A mammal that lived more than 200 million years ago named *Morganucodon* appears to be the earliest vertebrate to replace its teeth only a single time.[26] Some scientists interpret this to mean that it was also the first to rely on its mothers' milk in infancy. Over the following 50 million years new mammals appeared, including the ancestors of egg-laying platypuses as well as the two main groups of living mammals: *marsupials* and *placentals*.

Although we'll largely concentrate on placental mammals below—including humans and other primates—the enigmatic "duck-billed" platypus is worth a quick mention. Platypuses develop small vestigial teeth in their jaws during infancy, which are shed a few months later as they cease suckling and emerge from their breeding burrows.[27] Unique among mammals, the platypus has sacrificed its adult teeth to the development of an electrolocation system in its skull for finding prey underwater—which it then crushes with its hard, keratinized, toothless beak. They're fascinating to watch swimming in Australian waterways, as they can make quite a mess while blindly digging up the riverbed in search of a meal.

Our one-time replacement of teeth is another mammalian innovation, since nearly all other vertebrates replace their teeth again and again.[28] Tooth eruption and loss in most fish, reptiles, and amphibians is easier on the individual than you might imagine, since teeth tend to be more loosely bound to the jaw than those of mammals, who have tight, root-sized, bony sockets to hold them firmly in place. Like birds, young mammals have a period of rapid initial growth until they reach their final adult body size, while reptiles, amphibians, and fish continue growing slowly throughout life. This difference is paralleled in their dentitions, as non-mammalian vertebrates erupt increasingly larger teeth to fill their growing jaws continuously, whereas humans and other mammals typically replace their deciduous teeth with a final set as their skeletal development nears completion.

The two main groups of modern mammals that give birth to young directly are pretty distinct. Marsupials, found in the Americas and Australia, bear small offspring at an early developmental stage, tucking them into pockets or pouches on their body after birth. Many marsupials grow only one set of teeth, or replace just a single premolar tooth. The benefit of this odd pattern is unknown.[29] In contrast, placental mammals—named for the uterine tissue that feeds offspring as they grow internally, are born larger and more developed than marsupial mammals. Placentals typically share a pattern of tooth replacement with *Morganucodon* and other early mammals: a single set of deciduous incisors, canines, and premolars erupts early and is eventually followed by the molars and the rest of the permanent dentition.

Modern placental mammals have evolved a range of tooth sizes, shapes, and other variations of their ancestral pattern.[30] Rodents are well known for their single set of continuously growing incisors, allowing them to gnaw tough objects. Despite rapid loss through tooth wear, their sharpness and height is maintained by ongoing formation and eruption. Elephants also have continuously growing upper incisors, which have been modified into long tusks that assist with digging, marking trees, and fighting over mates. These tusks increase in length by about a half-foot each year, and can grow over 10 feet or 3 meters long! The rest of the elephant dentition consists of premolars and molars, which emerge in an unusual fashion. Elephants normally chew with a single premolar or molar tooth on each side of the jaw, and as this wears down another comes forward to take its place (figure

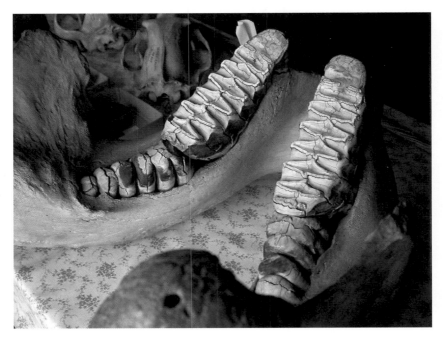

Figure 4.5
Elephant lower jaw showing a pair of erupted molar teeth (upper right)
and developing ones directly behind the large multi-crested teeth.
Specimen courtesy of the Humboldt Museum (Berlin).

4.5).[31] When the last molar is worn through, the elephant typically dies of starvation—a heartbreaking fate for such a majestic animal.

As we've seen, teeth play an important role in tracing the evolution of mammals. Ancient mammals had as many as 50 permanent teeth—still fewer than most fish, amphibians, and reptiles.[32] Early placental mammals typically had 44. Modern rodents and elephants have less; rodents sport 16–22 teeth, and elephants have only 26 all told. Tallying numbers and detailing tooth shapes allow paleontologists to reconstruct how mammals became more diverse throughout the reign of dinosaurs, as well as beyond another major mass extinction event 66 million years ago.[33]

The reasons for this extinction continue to be debated, but it's clear that drastic environmental changes occurred as a result of a large asteroid impact. Increased volcanic activity, falling sea levels, and cooling temperatures fundamentally altered the landscape. Many large-bodied organisms and dietary specialists died out, including 75% of all birds, 51% of all reptiles—famously including the dinosaurs—as well as 23% of all mammals. The animals that survived these changes rapidly diversified in next few million years, and numerous groups of modern mammals appeared, including our own evolutionary order, Primates. Some dub this era the "Age of Mammals"—a time when mammals became the primary large-bodied creatures roaming the land, filling the ecological niche previously occupied by dinosaurs and other hulking reptiles.

The Dawn of Primates

Primates appear to have found their footing much like mammals did 150 million years earlier. Shortly after the extinction event that induced the dinosaurs' demise, new types of primate-like mammals began to turn up in the forests of Asia, North America, and Europe.[34] Although the continents were in nearly the positions we know today, land bridges connected these regions—facilitating the rapid spread of small, possum-like creatures around the globe. They arose during a time when fruiting and flowering plants came to dominate the landscape. Like living primates, these mammals had a varied array of teeth, including pointy molars for crushing insect skeletons with deep basins for pulping fruit. We might consider them evolutionary cousins to primates, as they show fewer teeth and have large projecting incisors—features that seem too specialized to be found in the ancestors of modern primates. Yet their position in nature was soon to be overshadowed by the earliest true primates—masters of the trees with refined sensory systems to suit their new lifestyle.

The main reason for the rapid evolutionary success of primates is not entirely clear, and debates on the subject have been passionate over the past few decades.[35] Some anthropologists highlight how primate bodies are more agile than other tree-dwelling mammals. Specialized anatomical features enable some to forage for fruit, flowers, or young leaves in thin outer tree branches—a tropical niche otherwise only accessible to birds and bats. Early primates are believed to have had small bodies, as they had small teeth, which is a reasonable proxy for their weight. Primates who weigh less than 500 grams (about one pound) don't have large-enough digestive systems to extract much energy from leaves, and thus must rely on insects or other animal prey for dietary protein. This leads us to another theory for primates' success, as some scientists credit the forward-facing position of their eyes—allowing for overlapping visual fields and enhanced depth perception—that makes them effective insect predators and highly skilled leapers.

Yet another theory suggests that primates coevolved, or changed together through time, with tropical plants that bear flowers and fruit. We know that modern primates rely heavily on these for food, as well as the insects that are attracted to their flowers and fruit. Coevolution is hard to prove, but this quid pro quo system has been described as a main driver in the diversification of plants, pollinators, and seed-dispersers. Trees employ primates to spread seeds after they are swallowed, and harnessing these primate gardeners may have aided both groups. While anthropologists have gone back and forth about which scenario best explains primates' evolutionary success, their position as highly adept mammals is apparent to anyone who has visited a zoo or been on a wildlife safari.

The current fossil record indicates that the ancestors of living primates first appeared in the forests of Asia, moving into North America and Europe during a period of especially warm temperatures starting around 55 million years ago. It might be hard to imagine, but tropical forests and large bodies of water once covered North America and Europe. These were ideal environments for primates and other mammals to diversify, and the margins of ancient lakes and rivers are often the best places to find traces of life in the past. Animal bodies rapidly covered by mud or sand often preserve better than those that remain on dry land, where scavengers and bacteria race the elements to break down skeletal material—often leaving no trace behind. Much of the early primate material consists of isolated teeth or jaws, which are more resistant to destruction. Fortunately for those of us on the trail of our evolutionary journey, these fossils show subtle details that align them with living primates, including ourselves.

I had the privilege of collecting early primate fossils in the Great Divide Basin of Wyoming with my undergraduate advisor, Bob Anemone, during the summers of 1996 and 1997. Bob's paleontological expeditions are unforgettable, as we'd drive for miles off-road to reach prospective fossil sites and then camp out under the stars for weeks. Each day we'd rise early, stock up on water and peanut butter and jelly sandwiches, and head out into the hot dusty plains populated by pronghorn antelope and sagebrush. Bob would take his students to rock exposures of known geological ages and help us spot the characteristic sheen of fossil enamel (figure 4.6). I loved

Figure 4.6
Fossil primate jaw recovered from the Great Divide Basin in Wyoming.
Image credit: Robert Anemone.

being in the field—searching for sediments that contained fossils, crawl-ing around on my hands and knees looking for miniature jaws, and warily scanning anthills for the rare tooth in foot-high mounds of tiny pebbles. Ants often bring tiny teeth to the surface while excavating their burrows, and they aren't afraid to sting careless fossil hunters. At night my dreams were filled with visions of shiny black teeth peeking out from the beige soils and sands—more than 20 years later, this image still remains locked in my mind!

Humans and other primates living today are united by a number of anatomical specializations, including forward-facing eyes and dexterous hands with nails rather than claws—adaptations that make it possible to turn the page while reading a book, or pick up and identify tiny teeth formed millions of years ago. In this chapter, we've learned that the ancestral placental mammal had 44 teeth, and we know from chapter 2 that humans have 32 permanent teeth. Dental downsizing took place in stages: early primates originally lost four incisors and four premolars when they parted ways with their mammalian ancestors, followed by another four premolars when African and Asian (Old World) monkeys and apes diverged from other primates. Premolars were lost from the front of the primate jaw as the snout shortened, so you will hear biological anthropologists refer to the remaining two on each side as the third and fourth premolars, or P3 and P4.[36] The shift to shorter snouts may have come from relying on vision more than smell—part of a series of changes in the face, sense organs, and brain. In the following chapter, I'll explain how our hominin ancestors continued this makeover, reducing their moderately pointed, ape-like faces to a relatively flat human profile.

We are one of more than 600 primate species living today, which are now classified into two main groups: lemurs and lorises (technically called strepsirrhines), and tarsiers, monkeys, apes, and humans (technically called haplorrhines; figure 4.7). Some of these terms may be unfamiliar to you—although lemurs have become popular in Western culture thanks to their depiction in popular movies about their homeland, Madagascar. When most people think of a primate, they tend to picture a monkey or an ape—more recent evolutionary groups than the lemurs and lorises of Africa and Asia. These lesser-known varieties branched off the lineage that gave rise to monkeys, apes, and humans shortly after primates first appeared. They've had more than 50 million years to diversify, leading to incredible oddities such as the aye-aye, an elusive lemur that is active only at night. Aye-ayes look like a bat crossed with a possum—large ears frame its haunting eyes and pointed snout. From a dental perspective, they are one of the most specialized primates—evolving large, ever-growing incisors to gouge trees on a regular basis. Once they gnaw through the tough outer bark, they insert elongated middle fingers to retrieve energy-rich grubs from deep burrows in the trunk. Aye-ayes have reduced their dentition even more than humans, retaining only 20 teeth and configuring them like rodents.

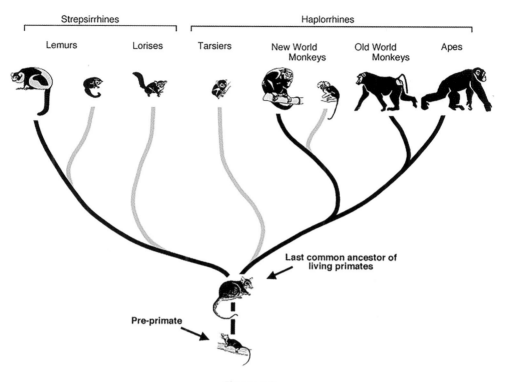

Figure 4.7

Evolutionary relationships among living primates. Reproduced from Christophe Soligo and Robert D. Martin, "Adaptive Origins of Primates Revisited," *Journal of Human Evolution* 50, no. 4 (April 2006): 414–430, with permission from Elsevier.

Other lemurs evolved curious adaptations such as a "tooth comb"—a group of lower incisors and canines that project horizontally relative to the jaw-bone (figure 4.8). These work as their name suggests—allowing lemurs to clean their fur much like humans comb their hair. Some lemurs also use these combs to gouge trees in order to feast on nutrient-rich sap. Primates have also found other uses for their choppers besides cutting and chewing, and you're likely to be surprised by some of the human behaviors discussed in chapter 9.

Current evidence suggests that the primate lineage that gave rise to monkeys, apes, and humans appeared in Asia nearly 55 million years ago.[37] During a subsequent period of environmental cooling that led to the extinction of North American and European primates, some members of this group took refuge in Africa. Not content to stay on one continent, a group

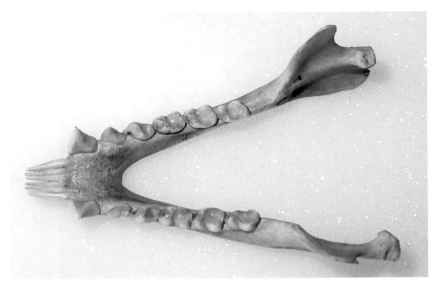

Figure 4.8
Lower jaw of a lemur from Madagascar showing the distinctive tooth comb (left).
Primate specimen courtesy of Thomas Koppe, University of Greifswald (Germany).

of them colonized South America by means of one of the most remarkable migration events in the history of mammalian evolution. We really don't understand how these "New World" monkeys made it across the Atlantic Ocean—which may have been slightly lower at the time, but was still a formidable barrier for land-dwelling animals. Since South America was likely isolated from the other continents, the most plausible explanation is that a large floating mass of vegetation broke off from Africa, carrying some very confused simian stowaways to a new shore. Modern New World monkeys include the aptly named howler monkeys, as well as capuchin monkeys—the iconic organ-grinder monkeys popular in the United States during the late nineteenth and early twentieth centuries. Wild capuchins are fascinating to watch—especially as they have figured out how to use rocks to crack open hard seeds, smartly protecting their teeth from such a demanding task.

The primates that remained behind in the Old World diversified into monkeys and apes about 23 million years ago. Both groups share the same dental formula, which is the same as ours: eight incisors, four canines, eight premolars, and 12 molars, or 32 teeth in total. However, apes and monkeys

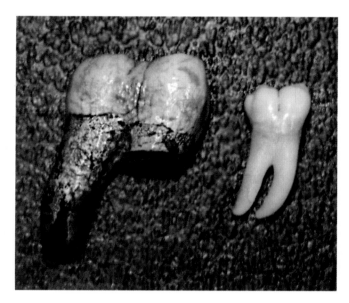

Figure 4.9
Comparison of lower molar of *Gigantopithecus*, the largest fossil
ape known, with a modern human molar (right). *Gigantopithecus*
fossil courtesy of the Senckenberg Museum (Frankfurt).

show important differences that make them easy to tell apart. Apes lost their
tails, grew bigger brains, and evolved relatively simple-looking teeth that
allowed for a generalized diet. Most apes have a preference for eating fruit,
but can do without when it's out of season. Old World monkeys routinely
consume leaves and other plant parts, as indicated by their taller molar cusps
and specialized digestive anatomy. Apes had the upper hand for millions
of years, spreading throughout Africa and eventually into Europe and Asia.
Although most fossil species are poorly known, their dental anatomy hints
at a variety of diets, habitats, and body sizes. Perhaps the most remarkable
is the Asian fossil genus known as *Gigantopithecus*—dubbed "the real King
Kong"—which have the largest teeth of any primate (figure 4.9). Extrapolations based on tooth size suggest that one species of *Gigantopithecus* may
have been taller than a human male and much heavier than a male gorilla.

Apes became rare by about 9–10 million years ago, ultimately disappear-
ing in Europe and parts of Africa. The direct ancestor of living African
apes and humans has proven difficult to identify, and there is a good deal

of debate about which kind of fossils are the most telling. Some scientists point to evidence from skulls or teeth to argue for similarities between fossil and living apes. But others prioritize evidence based on the anatomy below the head, which provides key information on how primates moved and which environments they may have favored. Part of the problem is that modern apes prefer environments that hamper the fossilization of skeletal remains. For example, there are very few fossils from the lineage leading to modern chimpanzees, and effectively none for gorillas, despite the fact that genetic studies have established that these groups have had distinct histories for more than 6–8 million years.[38]

As we'll see in the following chapter, those of us who study the lineage leading to modern humans have more evidence to work with, but paleoanthropologists continue to ponder how the numerous groups relate to one another and which groups were directly ancestral to humans living today. We've acquired many essential traits over the long course of vertebrate, mammalian, and primate evolution, and considering their origins and transitions helps us keep perspective on the ongoing human journey. Teeth continue to play an important role in debates about what makes us such unusual primates, especially since our development can be benchmarked from records of tooth growth. In the final chapter of this section, we'll review how development has changed over evolutionary time, reflecting a fascinating trend of elongation that distinguishes humans and great apes from all other primates.

5 From Humble Beginnings: Human Origins and Evolution

Those of us whose study our fossil ancestors and their relatives, called hominins, owe a great debt to Thomas Huxley and Charles Darwin. As I explained earlier, they courageously proposed that humans descended from primates and that our earliest ancestors would be found in Africa. These ideas were initially met with dismissal and accusations of religious blasphemy in nineteenth-century England, which held fast to ideas of a hierarchy of living beings created according to Judeo-Christian dogma. Humans were seen as the earthly pinnacle of divine creation rather than a result of gradual changes accumulated over millions of years. Great apes, known today as our closest living relatives, were largely unheard of in England during Darwin's youth. Moreover, what little was known of monkeys then wasn't terribly flattering, so it's no wonder that members of this formal Victorian-era culture had difficulty accepting that, yes—they too were primates!

The assertion that humans are related to primates can be traced further back to Carl Linnaeus, an eighteenth-century Swedish naturalist who developed the modern system of biological classification known as taxonomy. Many of us learned about his method for naming plants and animals through a consideration of our own place in nature. In 1758, Linnaeus famously classified humans as a unique species (*sapiens*) in the grouping known as a Genus (*Homo*). Since those two labels together form our Latin name, we are officially *Homo sapiens*, or "wise human beings."[1] He recognized our close similarity to primates by grouping humans in the order Primates; our affinity to other mammals by including us in the class Mammalia; and our broad relationship to other animals by placing us in the kingdom Animalia. Finer divisions of this scheme have since been adopted, but Linnaeus' basic logic remains in place for classifying both living and ancient forms of life, including plants.

The first hominin fossils were formally recognized and named around the time that Huxley and Darwin began writing about human evolution. *Homo neanderthalensis* was assigned to a partial skeleton recovered during quarrying work in the Neander Valley of Germany[2] (figure 5.1), and it remains a contentious classification due to recent evidence that Neanderthals and modern humans interbred. But we're getting ahead of ourselves, as this discovery didn't take place until Neanderthal DNA was sequenced in the twenty-first century. When the German fossil find was reported in 1856, scholars struggled to make sense of something that was like us in many respects—but also different in some important ways.[3] For example, the massive brow ridges framing its skullcap are not present in living people, yet the brain of this ancient species would have been as big as ours. We've already learned that this wasn't the first hominin discovered. A large-brained baby Neanderthal was unearthed in Belgium in the winter of 1829–1830, followed by another Neanderthal skull from Gibraltar in 1848—remains that were overlooked by scientists at the time. Stone tools had also been found that were unlike anything produced during recorded history, adding to the growing body of evidence that someone else lived in Europe a very long time ago.

The next ancient member of our genus was discovered far afield of the European Neanderthals: *Homo erectus* was found in Indonesia by a Dutch naturalist named Eugene Dubois.[4] Rather than being a complete surprise like the Neanderthals, its existence was predicted prior to its discovery. In 1887, Dubois took a job with the Dutch East Indies Army in order to find the "missing link" that would prove humans had evolved from apes. He started out on Sumatra, recovering a pair of human teeth that didn't seem terribly old, and paid them little attention. A historical side note is that these have recently been dated to 63,000–73,000 years ago, making them the most ancient remains of *Homo sapiens* from Indonesia.[5] By 1890, Dubois moved on to the island of Java with the hope of finding more suitable evidence of his ancient life form. During excavations in 1891 and 1892, his crew unearthed a large human-like thighbone, a skullcap with substantial brow ridges, and a few teeth—touching off an intellectual firestorm. Dubois struggled for years to convince the scientific community that these were from an ape-like individual that stood upright. Importantly, he viewed the adoption of an erect *bipedal*, or two-footed, posture as the first step toward becoming human.

Dubois' work marked a turning point in the formal study of humanity's origins.[6] He developed important methods to estimate the height of the

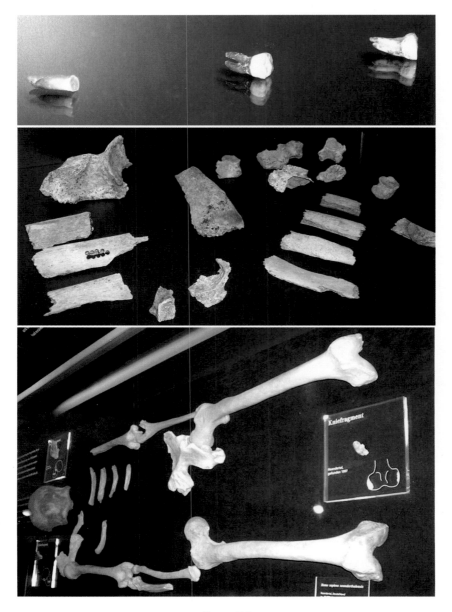

Figure 5.1
Remains of the iconic Neanderthal fossils from Germany. Drill holes in a bone
fragment were made for the recovery of ancient DNA. Images taken at the
Neanderthal Museum in Mettmann, Germany.

Homo erectus individual from the thighbone length, and estimated the size of its brain from the skull fragment—revealing that although this hominin stood as tall as a living person, its brain would have been only three-fourths as big as ours. A growing number of people began to accept that humans actually *had* evolved, and scientists started to formalize methods to study the accumulating skeletal and archeological evidence.

Dubois and his peers couldn't have anticipated the discoveries to come, including the ape-like Taung Child in 1924, or the even more primitive *Australopithecus afarensis* Lucy skeleton found 50 years later. These fossil finds permanently shifted the search for our ancient ancestors to the African continent. Paleoanthropologists today continue to make frequent paradigm-shifting discoveries—necessitating constant revisions of university textbooks on the subject! In the following pages, we'll trace how subsequent generations have tested and refined the "radical" ideas of Linnaeus, Huxley, Darwin, and Dubois—illuminating key changes in our anatomy and behavior that have led to our position as the sole survivor of a large clan of enigmatic hominins.

Who Went Where When?

Biological anthropologists have long endeavored to identify skeletal features that differ between humans and apes. We aim to determine whether newly discovered fossils should be grouped with the lineage leading to modern humans, or are more closely related to the great apes—chimpanzees, gorillas, and orangutans. This has proven to be more challenging than you might expect. Unlike apes, living humans have large brains that are well balanced over the spinal column, our legs are longer than our arms, and our fingers are straight and specialized for manual dexterity (figure 5.2). A number of additional specializations help us to stand and walk with relative ease, which can be seen in the spine, pelvis, and lower-limb bones. Unfortunately these traits did not appear in a single transition from an ape-like ancestor to our modern configuration. Quite the opposite—much of our defining anatomy has appeared piecemeal. This creates a problem, as it isn't clear where to draw the line between fossil apes and hominins. Is a single "human-like" specialization sufficient, or should we call something a hominin only if it shows all the specializations of living humans?[7] How about when only a small portion of a skeleton is recovered? Teeth have come to play a special

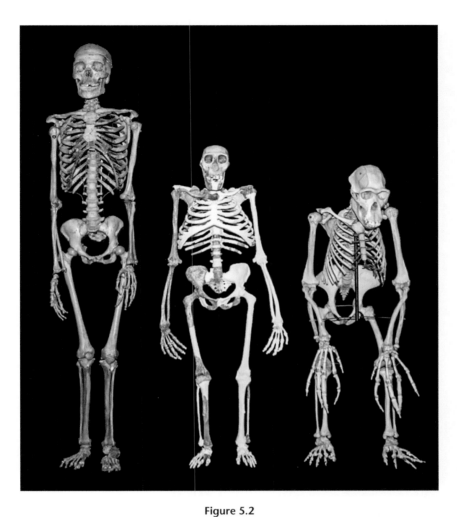

Figure 5.2
Comparison of modern human (left), australopithecine (middle), and chimpanzee (right) skeletons. Image from Lee R. Berger, "The Mosaic Nature of *Australopithecus sediba*," *Science* 340 (2013): 163. Reprinted with permission from the American Association for the Advancement of Science.

role in the resolution of this issue, although history has shown they can be misleading as well.

An example that has resulted in dramatic reinterpretations over the past century is the thickness of tooth enamel.[8] Until fairly recently, scholars used the degree of enamel thickness to classify fossil teeth as "human-like" or "ape-like." This is because people living today have thick enamel on their molars, while the African apes have relatively thin enamel. Yet the Asian orangutan—a more distantly related great ape—has enamel that is thicker than the African apes, complicating the division of enamel thickness into an ape-like thin condition and a human-like thick condition (figure 5.3).

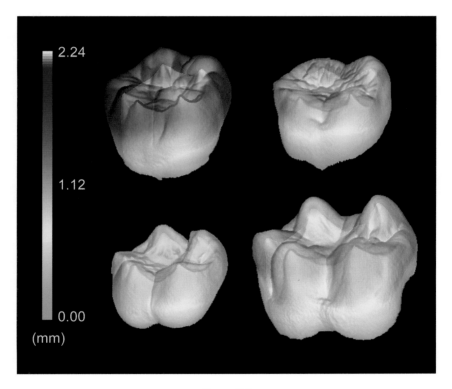

Figure 5.3
Enamel thickness in human and great ape molars. The color scale reflects the linear thickness of enamel in the tooth crown, in millimeters. Clockwise from top left: human, orangutan, gorilla, and chimpanzee first molars. Reprinted with permission from Reiko T. Kono, "Molar Enamel Thickness and Distribution Patterns in Extant Great Apes and Humans: New Insights Based on a 3-Dimensional Whole Crown Perspective," *Anthropological Science* 112 (2004): 121–146.

Knowledge of this detail emerged during efforts to study the famous paleo-anthropological hoax known as "Piltdown Man."[9] It all started shortly after Dubois' discovery of *Homo erectus*, which some scholars simply did not accept as a legitimate ancestral species. In 1912, a more palatable British missing link was presented to the scientific community—a reconstructed skull with ape-like teeth and room for a modern human brain. Parts of the cranium were supposedly discovered in a gravel pit south of London by an amateur collector named Charles Dawson, who convinced a paleontologist at the British Museum to aid in the search for additional material. Numerous debates followed about the evidence they presented, and the resulting drama served to increase public interest in the study of human evolution.

In 1918, the United States National Museum curator Gerrit Miller applied the new technique of X-ray imaging to the Piltdown jaw, publishing his findings in the first issue of the *American Journal of Physical Anthropology*.[10] What then was "physical anthropology" and now is "biological anthropology" includes paleoanthropology as well as the study of living humans and primates. By Miller's time it had become established as an international academic discipline, although even today biological anthropologists are less numerous than cultural anthropologists and archeologists.[11] Miller examined the internal structure of the Piltdown molars from the radiographs, arguing that the jaw was likely to have come from an ape rather than a human ancestor. Importantly, he was the first to show that thick enamel isn't unique to humans—since orangutans can have thick enamel too. Over the following years, a number of additional types of evidence led to the conclusion that the Piltdown skull was a hoax rather than a legitimate human ancestor. Miller had been right that the jaw came from an ape. Careful investigation has proven that parts of two artificially weathered modern human skulls were assembled into a single cranium, which was paired with a filed-down orangutan jaw. Charles Dawson died years before the truth of his elaborate deception came to light, perhaps mercifully, since it ultimately secured his infamy rather than the scholarly esteem he had hoped for.

In an ironic twist, Miller's insights about enamel thickness were overlooked more than 50 years later, when legitimate fossil material from India and Pakistan was misclassified as part of the human lineage.[12] Several jaws and teeth were used to define the now-invalid hominin genus "*Ramapithecus*," pushing our origins all the way back to 15 million years ago! The initial evidence was persuasive to many biological anthropologists, yet subsequent investigations over the following two decades proved that these fossils are not part of our ancestry. Confusion arose because the fragmentary "*Ramapithecus*" teeth looked rather human-like and had thick enamel. Yet Miller had already proven that thick molar enamel is not unique to humans. Moreover, genetic and fossil evidence points to an evolutionary divergence of human and chimpanzee lineages around 6–8 million years ago, making "*Ramapithecus*" too old to be part of our exclusive ancestry. Today most scientists believe that these fossils belong to the Asian ape *Sivapithecus*—a close relative of living orangutans. Humans and Asian apes have evolved similar dental features independently rather than from a common ancestor, representing another instance of parallel evolution. Miller summed it up perfectly: "It is evident that individual human teeth cannot always be identified by the characters of crown form supposed to be diagnostic."[13]

I've learned firsthand how challenging it can be to sort hominins from fossil apes, particularly when all you've got to work with are individual teeth found without the rest of the skeleton. Several years ago, I was invited to study the two molars Dubois' team had discovered in Java (figure 5.4), whose identity, believe it or not, scientists have continued to question for more than a century![14] As we've seen with "*Ramapithecus*," hominin and Asian ape molars can be easily confused from the outside, especially when

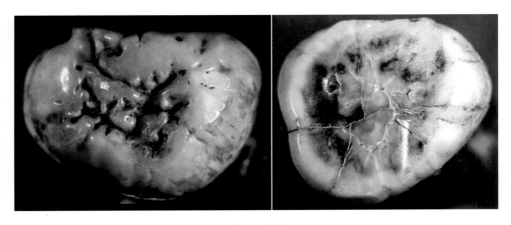

Figure 5.4
Fossil molars from Java recovered by Eugene Dubois' team. Reprinted with permission from Tanya M. Smith et al., "Taxonomic Assessment of the Trinil Molars Using Non-Destructive 3D Structural and Development Analysis," *PaleoAnthropology* (2009): 117–129.

the crowns wear down—blunting the original shape of cusps and fissures. I knew that we needed other lines of evidence to classify the Javan molars, since measurements of the size and shape of these teeth hadn't convinced previous generations of paleoanthropologists. So I enlisted Paul Tafforeau to help test Dubois' idea that they belonged to *Homo erectus*. We used the synchrotron imaging approach described in chapter 1 to study long-period growth lines inside the teeth, which showed 6–7 days between pairs of lines. This was similar to a definitive *Homo erectus* tooth from Java that had a 7-day rhythm. Importantly, Dubois' fossils showed lower values than fossil orangutan teeth, which have 9–12 days between long-period lines.[15] Their internal anatomy, as well as their root shape, was also similar to other *Homo erectus* teeth, leading us to conclude that the much-maligned naturalist had been correct.

Another longstanding debate in anthropology concerns how we define ourselves. Uncovering when, how, and why we became "human" is an important aim. Take tool use, for example, since Darwin and other early anthropologists pointed to this as a defining human trait. However, we now know that numerous animals use tools in the wild, including birds, sea otters, monkeys, and apes. Many people have learned about the chimpanzees of Gombe made famous by Jane Goodall and *National Geographic*, who fashion carefully selected twigs for "termite fishing." Recent studies have shown that chimpanzees use stone tools as well—prompting primatologists to partner with archeologists to investigate this unexpected behavior.[16] It seems that we can no longer consider traits like thick tooth enamel or tool use to be unique to humans and our ancestors. Defining ourselves has turned out to be much more complicated, as we'll see in the following section.

Human Evolution in Three Parts

More than 20 new hominin species have been added to the mix since Dubois' prescient expeditions, and a complex picture has replaced the classic idea of a single line of descent from ancient apes to a missing link to living people (figure 5.5). Today's educators often describe human evolution as a tree with numerous branches—many of which lead to tips that have gone extinct. My approach to explain the human journey is to liken it to a three-act play that has been running for several million years. Yet before we go too far with this analogy, it's important to keep in mind that

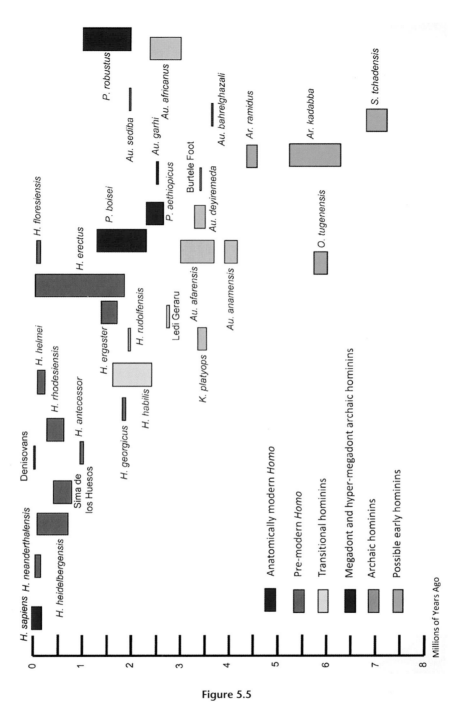

Figure 5.5

Fossil hominin species and unnamed fossil discoveries. Genus names have been abbreviated as follows: H.—*Homo*; P.—*Paranthropus*; Au.—*Australopithecus*; K.—*Kenyanthropus*; Ar.—*Ardipithecus*; O.—*Orrorin*; S.—*Sahelanthropus*. Reprinted with permission from Bernard Wood and Eve K. Boyle, "Hominin Taxic Diversity: Fact or Fantasy?" *Yearbook of Physical Anthropology* 159 (2016): S37–S78.

our history should not be seen as a dramatic progression that builds to a foregone conclusion. Humanity is a product of evolution—continual genetic changes driven by natural selection, mutation, and random changes termed genetic drift. Like all other living things, we've been shaped by past environmental dynamics, including climates, food sources, and pressures from predators and competitors.

"Scene One" takes place in the distant past of Africa, a continent inhabited by monkeys that roamed dense forests and sunny woodlands near rivers and lakes. The ancestors of modern African apes were there, too, although little is known about what they looked like or where they lived. Prehistoric records from equatorial Africa are harder to come by than those from other parts of the continent. Rainforests covered large swaths of this region, and acidic soils and rapid decomposition hinder fossilization of their occupants. As we'll see in the following pages, hominins that lived 4–7 million years ago have more in common with living apes than modern humans, and they likely continued to find frequent refuge in the trees.

Our understanding of this important period began in 1992, when Gen Suwa, a Japanese paleontologist, discovered a small molar tooth in an Ethiopian desert.[17] He was on an expedition with Ethiopian and American fossil hunters, and they had zeroed in on a promising region interspersed with ancient volcanic deposits. The first auspicious molar was quickly followed by the recovery of additional teeth, an upper arm bone, and part of a skull. The following year the team returned to the site, finding more skeletal fragments from a surprising ape-like hominin that was more ancient than *Australopithecus*, the oldest known fossils at the time.

Publication of the new genus *Ardipithecus* rocked the field of paleoanthropology—fueling another round of the debate over how hominins should be defined or recognized. The initial evidence was largely based on the shape of its teeth. A baby tooth—like most of the permanent teeth recovered—was smaller than those of *Australopithecus* and shaped more like chimpanzee teeth. Yet the discoverers were confident that this new genus was near the root of the human lineage. At the time, I was an undergraduate student in Bob Anemone's Human Evolution class, and I still have the simple evolutionary tree I drew for a class project connecting *Homo sapiens* and *Homo erectus* to *Australopithecus*, and ultimately to *Ardipithecus*.

It turns out that discoveries of a more ancient *Ardipithecus* species, as well as the 6- and 7-million-year-old hominin genera *Orrorin* and *Sahelanthropus*,

Table 5.1
Hominin Genera, Regions, and Approximate Time Periods[18]

Genus	Region	Dates (Millions of Years Ago)
Early hominins		
Sahelanthropus	Central Africa	7
Orrorin	East Africa	6
Ardipithecus	East Africa	6–4
Transition to classic australopithecines		
Australopithecus	Central, East, and South Africa	4–2
Kenyanthropus	East Africa	4–3
Paranthropus	East and South Africa	3–1
Transition to *Homo*		
Homo	Africa	3–present*
	Eurasia	2–present
	Europe	1–present

* Since the oldest fossil currently assigned to *Homo* at 2.8 million years is not universally recognized as a member of this genus, it is possible that more definitive material may bring this date to closer to 2 million years ago.

have complicated this picture (table 5.1).[19] Each of these has been reported to show specializations for an upright, bipedal posture. Yet there are few features that can be compared among these early hominins, making it difficult to determine whether or how they differ from one another. Canine teeth have been found from each group, and they are smaller than those of gorilla and chimpanzee males. This is a key detail, and one that Darwin predicted. He hypothesized that the robust canines used by male primates in aggressive interactions would get smaller in our ancestors, particularly with the use of weapons that could replace these dagger-like teeth. Darwin reasoned that as hominins became bipedal, males used their hands and arms for fighting and thus didn't need to use their teeth and jaws as much, leading to "a most striking and favourable change in his appearance."[20] We'll return to canine size in chapter 8, since this has served as an oft-discussed clue about the evolution of social systems.

"Scene Two" features the rise of the australopithecines, a diverse group that was quite successful from about 1–4 million years ago. The African landscape become more dry during this time, leading to a reduction in the forests that *Ardipithecus* had probably preferred and an expansion of savannahs and grasslands. Australopithecines spread out across Africa on foot—likely from an east African population that eventually moved west to modern-day Chad and south to modern-day South Africa. The oldest recognized species is *Australopithecus anamensis*, a presumed ancestor to the more well-known *Australopithecus afarensis*. The female skeleton dubbed "Lucy" in 1974 after the Beatles' song "Lucy in the Sky with Diamonds" epitomizes this younger species.[21] Nearly 30 years later, another member of this species captured the world's attention—an incredibly complete skeleton from Ethiopia heralded as "Lucy's Child" (figure 3 in this book's introduction). This fossil is of particular interest to me, since it's the most ancient infant known to science, and its teeth have important information about the evolution of our development locked inside—a topic we'll tackle in the following chapter.

In the beginning of this book I highlighted the first australopithecine discovered—the paradigm-shifting *Australopithecus africanus* (figure 4, introduction). In the decades after Raymond Dart's announcement of the remarkable Taung Child, additional jaws and teeth were discovered in South African underground caves, including those of other youngsters. Leopards or other predators may have captured them, bringing their defenseless meals up into trees that grew near the mouths of the caves. The iconic Taung Child may have been carried off and eaten by a large eagle, as scratches on its skull are similar to those of monkeys that have been preyed upon by African crowned eagles.[22] Hominin bones fell into caves or were washed into openings in the ground, where they were jumbled together and eventually cemented into dense limestone. Once scientists began to dissolve and chip away the concrete-like matrix that protected these fossils, it became clear that *Australopithecus africanus* wasn't alone. This species was only one example of a diverse radiation of australopithecines, which included several striking forms with massive faces and expanded premolars and molars. These "robust" master-chewers have their own genus, *Paranthropus*, to distinguish them from more lightly built, or "gracile," *Australopithecus* species.

Robust australopithecines are noteworthy because of their extremely thick-enameled teeth and heavily reinforced heads, which have a less pronounced snout than apes and earlier hominins. We'll revisit these odd hominins in chapter 7, since their unique skull anatomy and unusual tooth wear have led to confusion over what they were eating. Actually, the distinction between robust and gracile australopithecines may be somewhat misleading, since we don't know much about how they compare below the head. Bones are often too jumbled and fragmentary in most South African caves and east African sediments to make out what the bodies of *Paranthropus* looked like.[23]

A rare pair of associated skeletons was found in 2008, representing a small-toothed, small-brained hominin known as *Australopithecus sediba*. This gracile South African species shows an unexpected mix of ape-like and human-like traits in the hands and feet (figure 5.2).[24] We think that differences in australopithecine locomotor styles, facial structure, tooth size, and enamel thickness are likely to be adaptive responses to new dietary niches—not unlike the diversification of Darwin's finches in the Galápagos. Ultimately, this variation may have been driven by environmental change as well as competition among different hominins.

Several other australopithecines have been formally described, and the list of species continues to grow every few years. Some paleoanthropologists have criticized the tendency to create new species based on the recovery of a single fossil jawbone, since skull morphology can vary greatly between the sexes of great ape species. When compared to females, male apes typically have larger teeth and bones, with marked bumps and ridges for the attachment of powerful muscles. It's possible that scientists could mistake two fossils that show normal male-female variation for two separate species, particularly when the sex of the fossils is unknown. Another complication is that lower jaws look a lot alike, making it hard to separate fossil species from each another. In the few cases in which isolated jaws have been designated as new species, it's difficult to know what the missing cranium looked like, and impossible to infer the shape of the rest of the skeleton. This conundrum can only be resolved when a complete skeleton with a matching lower jaw turns up, which unfortunately happens about as often as a blue moon!

Over the past few years, there has been another major shift in our understanding of human evolution. Returning to the question of tool use, it turns out that more than 3 million years ago, australopithecines with ape-sized brains began making and using simple stone tools.[25] As I noted earlier, until recently this skill had been considered a landmark innovation of our own genus *Homo*. You may have heard about Louis and Mary Leakey's work at Olduvai Gorge in Tanzania, where they found a fragmentary skull and a partial skeleton with a set of hand bones in the 1960s. The strongly built bone at the tip of the thumb would have supported precise and powerful gripping abilities, leading to its classification as "the handy man," or *Homo habilis*.[26]

Furthermore, the fossils were recovered in close association with primitive stone tools, which the Leakeys believed were made by this species. Since then, many have described the transition from australopithecines to early *Homo* as a shift in cognitive abilities and tool-making strategies that improved dietary quality. Yet this is no longer so straightforward, since earlier hominins and even chimpanzees can make and use tools. Paleoanthropologists continue to wrestle with how to best define our genus, as well as whether these small-brained australopithecine-like "transition fossils" should be included.[27]

Diving into the details of our own genus brings us to "Scene Three"—a period of increasingly rapid and unpredictable climate change that began 2–3 million years ago. Some scientists feel that the resulting ecological insta-bility sparked the evolutionary origin of our own genus, since survival now demanded a new degree of behavioral flexibility and innovation.[28] Unfor-tunately it is difficult to test this idea, in part since the early fossil record of *Homo* is quite meager until about 2 million years ago. Ancient jaws and teeth from two Ethiopian sites may be the oldest evidence at 2.3 and 2.8 million years ago.[29] Yet how can we be sure they are not australopithecines? Paleoanthropologists typically use tooth size and shape to classify homi-nins into the major groups—a necessary step, since multiple species lived side by side in South and East Africa for more than a million years. *Paran-thropus* sported extremely large premolar and molar teeth, early *Homo* had much smaller teeth, and *Australopithecus* tends to fall in between.[30] Yet the recent discovery of the small-toothed *Australopithecus sediba* has challenged these divisions, as have similarities in jaw shape between this species and the putative early *Homo* mandible from 2.8 million years ago.[31]

Setting aside the question of which fragmentary fossil is the earliest member of *Homo*, we can consider that Dubois' missing link, *Homo erec-tus,* shares enough in common with living humans to be grouped in the same genus.[32] Adventurous *Homo erectus* populations are the first hominins known to have emigrated from Africa, reaching the today's Republic of Georgia in Eurasia 1.8 million years ago and Southeast Asia shortly there-after. The extraordinary discovery of these immigrants in the sediments below a medieval Eurasian village revealed their similarity with African *Homo* fossils, including the retention of medium-sized brains and use of simple tools.[33] Over time, *Homo erectus'* brains became larger while their faces and teeth began to shrink. I'll return to the subject of tooth reduction below, when we consider how human teeth have also gotten smaller over the course of our recent evolution. *Homo erectus* also was the first to evolve a modern human-like body below the head—an important anatomical reorganization that has been related to new efficiencies in locomotion, hunting, and food processing.[34]

The path of descent from *Homo erectus* to *Homo sapiens* isn't as straightforward as my undergraduate Human Evolution class drawing attempted to convey. I knew at the time that there had been considerable disagreement about whether all modern humans descended from a source population in Africa, or whether various hominins in Europe, Asia, and Africa gave rise to the people living there today. Several species of *Homo* have walked the earth during the past million years, including *Homo neanderthalensis*, which—with brains similar in size to people living today—seemed to some like a plausible European ancestor. Moreover, unexpected populations of *Homo* have turned up in places like Siberia, including the "Denisovans"—close relatives of the Neanderthals represented by a few teeth and a finger bone that have yielded copious ancient DNA. To muddle matters further, the new small-brained species *Homo floresiensis* and *Homo naledi* prove that primitive features can be retained by some species for millions of years after they evolved from an australopithecine ancestor. Making sense of our line of evolutionary descent has only gotten more complicated over the past few decades.

Today most paleoanthropologists and evolutionary geneticists favor the idea that these "archaic" groups were replaced by a population of *Homo sapiens* that left Africa roughly 50,000–75,000 years ago.[35] This was apparently a somewhat dramatic transition in places where groups overlapped. Scientists have recently detected that modern humans not only encountered other species but also interbred with some of them! The details of these "foreign exchanges" are still coming into focus as ancient DNA retrieval becomes more sophisticated and as our understanding of human genetic variation deepens.

The current consensus is that all non-African populations living today carry a small percentage of Neanderthal DNA in their genomes, and Asians and Australasians also have a similar genetic contribution from Denisovans. This implies that interbreeding events with these hominins occurred as our species left Africa and began to settle in Eurasia and the Indo-Pacific. A few years ago I undertook my own genetic test to enliven the Human Evolution course I was teaching at Harvard University. Apparently my DNA shows a 3.1% contribution from Neanderthals—an amount that exceeds

97% of people of European descent! My students found this amusing, since I was pretty heavily engrossed in the study of Neanderthals at that time.

Now that the Neanderthal genome has been deciphered, it is possible to use statistical methods to estimate when *Homo sapiens* and *Homo neanderthalensis* diverged. Setting aside the recent interbreeding events, these lineages went their separate ways genetically between 550,000–765,000 years ago—making the Neanderthals our evolutionary cousins rather than our ancestors.[36] This means that our own species has had a longer history on earth than the fossil record alone suggests. Identifying ancient *Homo sapiens* individuals isn't as easy as you might think. One surprising trait that unites living and fossil members of our species is the presence of a chin. We don't really know what it's good for, beyond creating an attractive facial profile—but chins are not found in any other hominin. Several large-brained ancestral candidates lived during the last 600,000 years in Europe and Africa, but they don't exactly look like people living today, and we have few preserved lower jaws that would show whether they had chins.[37] It's not until about 300,000 years ago that subtle aspects of our faces, jaws, and teeth appear to distinguish fossil *Homo sapiens* from other contemporaneous archaic groups.[38] My research team has discovered that slow dental development and thick tooth enamel also set *Homo sapiens* apart from Neanderthals.[39] When these disparate lines of evidence are knit together, it sure looks like the curious anatomy that makes us who we are today appeared bit by bit, rather than all at once.

Homo sapiens and Dental Reduction

You may have gotten the impression that paleoanthropologists don't agree on much, which is only partially true. A good example of unanimity is the recognition that teeth have gotten smaller over the course of human evolution, although why this is so remains an open question.[40] Remember that the first hominins had smaller canines than most great apes—and this reduction continued throughout the subsequent diversification of australopithecines. Incisor size has fluctuated through time, ultimately decreasing in *Homo sapiens*. Similarly, small premolars and molars are a hallmark of early *Homo* individuals relative to *Paranthropus* and *Australopithecus*—a decrease that has further continued in our own species. In the final section of this book, we'll delve deeper into dietary, social, and cultural changes in our evolutionary history, but I'll give a brief summary of possible reasons for tooth shrinkage below.

One influential theory for why our teeth and faces initially got smaller relates to the use of fire, which can soften food and tenderize meat during cooking—potentially decreasing the need for forceful or prolonged chewing.[41] Although most of us don't cook over open fires too often, our modern system of food preparation relies heavily on heating, including frying, roasting, boiling, and microwaving. However, here's a place to exercise caution before drawing conclusions: just because two things occur together doesn't mean that one causes the other. While it seems logical to imagine that routine cooking led our teeth to shrink, evidence for the controlled use of fire earlier than 1 million years ago is in short supply.[42] Archeologists who investigate cultural behavior in Europe suggest that it wasn't until about 350,000 years ago that fire use became routine and widespread.[43] This is nearly two million years after our premolar and molar teeth began to get smaller, suggesting that other factors must have been at play.

Some have hypothesized that the use of tools and the consumption of meat led to the initial reduction of teeth in early *Homo*. An experiment by my former Harvard colleagues, Katie Zink and Daniel Lieberman, measured the number of chews and the amount of force needed to eat sliced, pounded, or roasted goat meat and root vegetables.[44] They found that slicing raw meat makes it easier to chew than roasting or pounding it. Similarly, pounding root vegetables makes them easier to ingest than eating them whole or sliced. Zink and Lieberman surmised that we don't have to invoke cooking to explain tooth reduction, arguing that routine tool use made chewing more efficient in *Homo erectus* than in australopithecines. But as we've seen, recent discoveries have suggested that australopithecines cut and broke animal bones with tools more than 3 million years ago. Food processing significantly predates *Homo erectus*, casting some doubt on the idea that their teeth became smaller because tool use relaxed the need for large choppers. Current evidence for the timing of both cooking and tool use doesn't align with changes detected in the record of fossil teeth. In all fairness, it's very difficult to determine how commonplace these behaviors were millions of years ago, so we would be wise to keep an open mind.

In a paper delivered to the First International Symposium on Dental Morphology in 1965—a key event for tooth aficionados that now occurs every three years—C. Loring Brace urged his fellow dental anthropologists to focus on measuring the teeth of modern humans.[45] Numerous research-ers have since heeded his advice—demonstrating that our teeth have con-tinued to evolve since *Homo sapiens* first appeared. Twenty years after publishing his "call to arms," Brace attempted to quantify the speed of this change, suggesting that over the last 10,000 years the size of the chewing surface has declined an average of 1% every 1,000 years. While there are picky reasons to quibble with his methods, this general trend has been veri-fied by others.[46] Comparisons of pre- and post-agricultural populations are particularly striking. Recall that the invention of agriculture, including the domestication of animals, occurred independently in several regions across the globe starting 10,000–15,000 years ago. As discussed in chapter 3, early agriculturalists often experienced more nutritional stress and developmen-tal defects than preceding hunter-gather populations.[47] Diets became more predictable and less varied. Living in close quarters with animals and large populations may have led to the spread of disease and inequity through social stratification. As these permanent settlements grew, humans became somewhat shorter and less well muscled, and our shrinking teeth were apparently part of this change.

What is less clear is *why* our teeth have continued to get smaller dur-ing and after this period. Numerous suggestions have been made, which largely consist of linking this trend with technological changes such as the development and use of cookware, spear throwers, or seed-grinding imple-ments.[48] Brace built on Darwin's line of reasoning about canine reduction, proposing a rather controversial idea: when natural selection for large teeth was relaxed due to a change in food processing, he suggested, genetic muta-tions were more likely to occur, ultimately leading to a decrease in tooth size.[49] Alternative explanations include the possibility that teeth became smaller simply because our faces reduced in size, or that building smaller teeth required less metabolic energy and dietary minerals. Others have sug-gested that natural selection led to smaller teeth in recent humans to keep them alive, as large-toothed individuals are more likely to have crowding, cavities, or potentially fatal impacted teeth.[50] While superficially compelling, especially to those of us who've suffered with impacted wisdom teeth, this idea doesn't account for why tooth size began to decrease thousands of years before impaction became common.

It turns out that humans aren't the only primates to have undergone den-tal reduction over time. Teeth also got smaller in one of my favorite primates, orangutans—and we can be sure that this wasn't due to advances in their cooking or hunting technology! Orangutans have an extensive fossil record in Asia, unlike the other great apes. Porcupines fortuitously brought the teeth of deceased individuals into caves, where they'd gnaw away the bone and root, discarding the enamel-covered crowns. Over thousands of years, these crowns would become embedded in the floor sediments that eventually hardened into cement-like time capsules. Scientists have recovered and measured thousands of fossil orangutan molars from Southeast Asia. When compared with modern orangutans living in Borneo and Sumatra, their fossil predecessors typically had much larger teeth.[51] I've determined that orangutans show a different pattern of dental tissue reduction than modern humans. The molars of orangutans became smaller through an equal loss of enamel and dentine, whereas humans have lost more proportionately more dentine than enamel over the evolution of our species.

Changes like this may point to distinct evolutionary processes. Modern-day orangutans are found on large islands that were once connected to mainland Asia but have since been separated for many thousands of years. When populations of mammals remain genetically isolated for long periods of time on islands, they often get smaller than their mainland progenitors. Orangutans' shrinking teeth may have been part of a general trend of size reduction that affected their whole body.[52] This dwarfing effect is famously evident in the 80,000- to 700,000-year-old hominin found on Flores, an Indonesian island.[53] *Homo floresiensis*—dubbed "the Hobbit"—is estimated to have stood at only about 3 feet tall with a small head and little teeth. While we aren't sure which species *Homo floresiensis* descended from, the leading mainland candidates *Homo habilis* and *Homo erectus* both had significantly larger bodies and teeth. Thus it's not unreasonable to hypothesize that geographic isolation on Indonesian islands may have played a part in orangutan dental reduction as well.

Homo sapiens' teeth have gotten smaller without such genetic segregation, and our jaws and facial skeletons have actually reduced more than other aspects of our heads or bodies. We've already discussed how some scientists have linked tooth reduction in our ancestors to the consumption of foods that have been cut, pounded, or cooked, which require lower chewing forces. I'm hopeful that paleoanthropologists will continue to search for explanations of *how* this occurred in parallel with reasons for *why* it happened.[54] For the time being, it seems reasonable to assume that dental reduction occurred at different points in time for different reasons, and in some cases it may have simply occurred by chance. Future work that integrates observations of fossil and archeological populations with modern experimental studies may one day solve this mystery.

Numerous changes over the past several million years have shaped who we are today, with large brains, small teeth and faces, and upright bodies. Another approach to the story of our own evolution is to begin in the present and look backward. Biological anthropologists often ask questions like: What makes us unique among the living primates? When did these characteristics appear? Why might they have evolved? As we've seen, evidence for our special way of toddling around appeared during the first stage of the human journey some 6–7 million years ago. By reorienting bones and muscles to support bipedal walking, hominins covered distances on the ground more efficiently than living apes—freeing their hands to carry tools, weapons, and food. Similarly, changes in brain size and structure ultimately facilitated more complex social networks, tool production, and food procurement, as well as language—a key innovation whose origins are not well understood.

Another suite of unique human characteristics—our long childhood, late maturation, and long life span—likely relates to having large brains as well. Importantly, these stages are also connected to our skeletal development, especially our teeth. While most life-history stages do not leave behind direct traces in the fossil record, teeth have been pivotal in efforts to trace the evolution of human development. As I've hinted at throughout this book, my colleagues and I use the time lines in teeth to understand how hominin species and other primates developed, evolved, and behaved. In the following chapter, we will review evidence that has helped to settle the controversy about how "modern" these extinct hominins were.

6   Evolutionary Perspectives on Human Growth and Development

Even when making allowances for our natural bias, we can say that humans are unusual primates. We have long pregnancies that result in the birth of helpless babies. Our infants stop nursing at young ages, but have long childhoods with a late growth spurt. Women bear their first child at relatively late ages, but can have kids as often as every year, stopping long before the end of life. *Life history*—or the way animals use energy for growth, reproduction, and staying healthy—is fundamental to biology and behavior. Studying this pattern may help to explain how natural selection facilitated the remarkable success of our species, while our close relatives are on the brink of extinction.

Exploring the evolution of human development from the limited fossil record is challenging. Until recently, our conception of the evolutionary transformation from an ape-like ancestor to the modern condition has relied on the assumption that extinct species had similar patterns of growth to either living apes or humans. Yet, as we've seen in the previous chapter, our ancestors evolved in a range of environments and possessed unique anatomical specializations. A growing body of evidence suggests most hominins were unlike any primate species living today. Let's see how modern science is uniting remote studies of wild primates, hard-tissue biology, and X-ray imaging to probe the fossil record in ways Darwin could not have imagined.

The Primate Life Cycle and Molar Eruption

Human life history differs from that of our closest living relative, the chimpanzee, in several ways (table 6.1). Anthropologists have struggled to explain this variation, particularly since relatively abundant australopithecine and

Table 6.1
Average Life-History Variables for Female Primates (in Years)[1]

Primate	Length of Pregnancy	End of Nursing (Weaning)	Age at First Birth	Birth Spacing	Maximum Life Span
Baboon (wild)	0.5	1.2	6.5	2.1	28.4
Chimpanzee (wild)	0.6	5.3	13.2	5.9	56.1
Human (non-industrial)	0.7	2.4	19.5	3.2	87.3

Neanderthal juveniles continue to stimulate questions about when and in which species our various modern attributes originated.[2] These inquiries have fueled studies of how primates and other mammals budget their metabolic energy, particularly while they are growing up. Budgets typically require compromise, and the metabolic budget is no different—reflecting tradeoffs in how energy is used. For example, when a young person's health is seriously compromised, the body may slow skeletal growth or curtail reproductive functions to bolster the immune system. Environmental factors such as the likelihood of having consistent and reliable food sources, or the risk of being preyed upon, also influence how organisms budget their energy. Mice—an easy meal for raptors, snakes, and carnivorous mammals—grow up quickly and have offspring as fast as possible, while elephants raise a small numbers of calves much more slowly. Elephants can afford to take their time, since they typically have a stable food supply and few natural predators thanks to their large size.

Understanding the life history of a species also tells us about its ability to respond to change. Rodents are more adept at colonizing new environments than many other mammals due to their rapid life cycle. Primate mothers, including humans, fall on the other end of this developmental spectrum, raising only a few slow-growing offspring during their reproductive years. Great apes are some of the slowest-developing mammals on the planet, which makes them especially vulnerable to extinction. Females don't begin reproducing until they are in their teen years, and they don't often live past their 30s or 40s.[3] To make matters worse, ape infants are more helpless than those of monkeys, requiring direct care for far longer. With intervals of five or more years between births, females only have a few offspring over their lives—the key marker of success from the vantage point of evolution. Orangutan mothers may nurse each offspring for more than

8 years—the latest weaning of any mammal![4] Sadly, orangutans are critically endangered; frequent forest fires, hunting, and deforestation driven by the palm oil industry are pushing them toward the brink of extinction.

As we discussed in chapter 4, animal development can vary greatly. Some fend for themselves after hatching from eggs, while others cling to their mothers for their first few months or even years of life. In this latter case, mothers benefit from being part of a social group, where they may leverage the protection of males and the support of other females. For example, some monkeys share infant care, which is thought to provide a respite for mothers and important babysitting practice for young females. Cooperative care is particularly common in monkey species where adult females remain in the group, so infants are likely to be cared for by sisters, aunts, and grandmothers.[5] These species often have strong dominance hierarchies, in which the relatives of the "alpha female" fare better than those of lower-ranking females. Great ape mothers don't benefit from the close-knit support seen in some monkey and lemur species. For example, chimpanzees tend to live in groups of unrelated adult females, while orangutan mothers are on their own save for occasional visits by solitary males.

Mammalian mothers face a trade-off between weaning each child quickly in order to prepare for the next pregnancy or nursing for a longer period of time to boost the survival of each offspring. Wild chimpanzee mothers who transition to bearing their next offspring rapidly do so at the expense of their previous one.[6] Quickly weaned infants remain small throughout their juvenile period in comparison to the offspring of mothers who coddled them for longer. While this doesn't seem ideal from the infant's perspective, chimpanzee mothers who can replenish their energy stores and bear young rapidly may leave more descendants than those who aren't in such a rush. Humans may have found the ideal solution to this dilemma by weaning their infants at an early age and then leveraging alternative energy sources to further their little ones' development. Mothers in traditional societies turn to their families—particularly older children and grandparents—to share the burden of childcare.[7]

Human mothers can easily double the reproductive output of most apes, meaning that our populations can increase much faster.[8] Great apes have been on the decline so rapidly that species such as the mountain gorilla number approximately 1,000 individuals in total, while the human population surpassed 7.5 billion in 2017. Adding to this is the fact that life expectancy in Western countries has doubled over the last few centuries[9]—mothers and *their* mothers are living longer than ever before. Cultural developments such as hygienic food production, sanitation, and medical care have certainly played a role in humanity's success. Importantly, our reproductive advantage over great apes is true even in traditional small-scale societies without these comforts.[10] In addition to our cultural advantage, human sociality has played an important role in our recent colonization of the globe. Anthropologists are well positioned to investigate these consequential trends.

We first met the Taung Child, *Australopithecus africanus,* at the beginning of this book. Recall that Raymond Dart's description included a discussion of its first molar eruption. At the time, it wasn't clear when this youngster's first molars emerged into the mouth, nor what that might have meant about its overall growth and development. A decade later, the biological anthropologist Adolf Schultz pieced together data on tooth eruption ages in living primates, proposing an influential model of the evolution of human development.[11] Schultz related the timing of tooth eruption to specific developmental stages: the eruption of the first permanent tooth marked the end of infancy, while the eruption of the last tooth marked the end of the juvenile period, or beginning of adulthood. He compared earlier-evolving primates, such as lemurs, and more recent ones, including chimpanzees and humans—noting that tooth eruption ages had gotten later as the overall development span had elongated over the course of evolution. Schultz reasoned that "early man" would have been intermediate between chimpanzees and humans, with an infant period more like that of humans, but a lifespan more like chimpanzees.

Anthropologist Holly Smith had the good sense to follow up on Schultz's hypothesis, and in 1989 she expanded his comparison of primate dental development and life history.[12] Smith found that the age of first molar eruption is strongly linked with weaning age and first-reproduction age in female primates. Primates with late-erupting first molars are weaned at older ages and begin reproducing later, too. She also found similar associations between third molar eruption and these life-history traits. Smith then extrapolated from these trends in living primates to infer life history in the human past. For example, she reasoned that since first molar eruption and weaning occurred around the same age in 14 species, weaning age could be determined for fossil hominins by aging individuals who died while erupting their first molars. This exciting possibility inspired numerous studies of hominin tooth growth over the following decades—the subject of the following section.

An association between first molar eruption and weaning makes sense, since primates need to have a working set of teeth to eat solid foods once they're no longer able to tap their mothers for milk. Smith reasoned that the addition of permanent first molars to rows of small deciduous teeth helped nonhuman primates with this dietary transition. However, some have started to question whether first molar eruption is an accurate predictor of weaning in all primates.[13] For example, first molars in orangutans, lowland gorillas, and chimpanzees appear a year or more before they are weaned, while humans cease nursing several years before their first molars are in place.

I decided to learn more by teaming up with some chimpanzee experts from Harvard.[14] They had recruited several photographers to record images of wild chimpanzees' mouths while yawning and playing (figure 6.1). We were able to use these pictures to see how molar eruption matched up with feeding behavior and reproductive histories. Surprisingly, we found that first molar eruption occurred long after they first started eating solid foods, but well before they stopped nursing completely. For example, the female infant in figure 6.1 erupted both lower first molars shortly before this photograph was taken at 3.1 years of age. She continued to nurse until 4.3–4.4 years of age, longer than expected based on Smith's study of primate dental development and weaning ages. We reasoned that first molar emergence in chimpanzees may be pegged to the age when they begin to feed on solid foods as often as the adults around them, which happens before they stop nursing completely.[15] Until this point, infants rely heavily on their mothers for nutritional support as they grow their tiny jaws and teeth, and rarely survive on their own if orphaned.

After poring over thousands of photographs of chimpanzee teeth, I had the pleasure of finally coming face to face with my subjects in Kibale National Park, Uganda (figure 6.2). This included meeting the unforgettable male named Bud, whose third molars emerged just after 16 years of age— an exceptionally late age that may mean they had been partially impacted. Another key result from our study is the observation that two young females erupted their third molars several months to several years before the birth of their first offspring. This runs counter to the idea that third molar eruption and first reproduction occur at the same age, as reported for other primates. You may wonder why primate-wide data sets fail to predict developmental

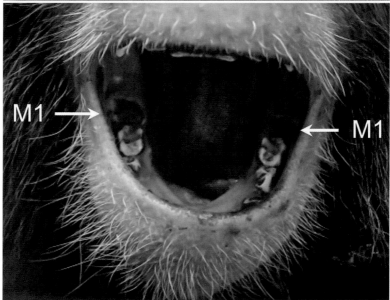

Figure 6.1
Young female chimpanzee erupting her first permanent molar (M1).
Reproduced from Tanya M. Smith et al., "First Molar Eruption, Weaning,
and Life History in Living Wild Chimpanzees," *Proceedings of the National
Academy of Sciences USA* 110 (2013): 2787–2791.

Figure 6.2
The author in Uganda in 2014 watching Bud, a male chimpanzee.
Image credit: Zarin Machanda.

milestones in chimpanzees, when they seemed to work well for other primates. Most of the life-history data available to Smith for her 1989 study came from captive animals—which we now know tend to have accelerated life histories in comparison to their wild counterparts.[16] Captive primates often wean earlier and begin reproducing at younger ages than those in the wild. Unfortunately, this means that while Smith's work sparked an exciting and important line of inquiry for the field of paleoanthropology, we ultimately cannot use molar emergence ages to predict traits such as the age of weaning or first reproduction in fossil hominins. But before you lose hope—let's explore a few other ways to learn when our unique life history appeared.

Human Dental Development through Time

The Taung Child wasn't the first fossil child discovered, but it was the first to gain widespread international fame. Dart noted that human children with recently erupted first molars like those of Taung would be about 6 years old. Most of his peers rejected this comparison and emphasized its ape-like anatomical similarities, suggesting that its rate of development would be more rapid, like that of apes.[17] This would have made Taung around 3 years old when it died. Since this time, numerous scientists have estimated the ages of fossil juveniles from the growth patterns of living humans and apes. This approach assumes that extinct species actually grew up like these modern groups—despite millions of years of evolutionary divergence. Yet we know today that each of the great apes develop in unique ways, and they have been influenced by different environmental and genetic changes during their evolution. Similarly, modern humans show subtle developmental variation in different populations, complicating efforts to apply growth standards from one group to interpret the fossil record.[18] Finally, there is the problem of circular reasoning. If a fossil child is aged according to human or ape developmental models, it isn't possible to really know how similar it is to either group.

The solution to this problem—a direct method to determine how old an immature fossil was when it died—is simple but not easy. In chapters 2 and 3, we discussed how dental development is a better predictor of a child's age than other measures like height, weight, or the size of individual bones. This is because the body protects tooth growth from minor fluctuations in nutrition or health during childhood—teeth, unlike bones, can't "catch up" later. As you'll recall, the microscopic biological rhythms in teeth aren't halted by most disruptions, recording the days of our lives from before we are born until our teeth finish forming. Studying tooth microstructure is the only independent path to explore maturation rates in fossil species, since it avoids the need to extrapolate growth in the past from populations living today.

This area of paleoanthropological research got its start in the 1980s, when two young scientists decided to count microscopic growth lines on the outsides of fossil teeth.[19] Timothy Bromage and Christopher Dean were the first to estimate the age of juvenile hominins from biological rhythms in teeth. Their work provided important evidence that *Australopithecus, Paranthropus,* and early *Homo* individuals formed their teeth at a pace more like great apes than modern humans. Although the details of their initial study have been refined over the last 30 years, numerous follow-up studies, including my own, have confirmed that australopithecines cannot be accurately aged from human growth standards, including those we discussed in chapter 2. They were unlikely to have shared our life history, since the timing of tooth calcification, speed of root formation, and first molar eruption ages suggest an earlier onset of adulthood than in people living today.

What about the Taung Child? Age estimates between 3 and 6 years have been the norm since 1925, and they've continued to shift within this range as methods of study improve.[20] In 1985, Bromage argued that Taung died shortly after it turned 3 years old, since it was very similar to a juvenile known as Sts 24 that was estimated to have died at 3.3 years of age. Twenty years later, a more detailed study of Taung's dental development by Bromage and colleagues revised this to between 3.7–3.9 years of age. In 2015, my research team conducted a study of Sts 24 and several additional australopithecines, finding that Sts 24 was a year older when it died than Bromage and Dean originally estimated. Three juveniles from Taung's species showed a range of developmental patterns. One appeared to grow its teeth more slowly than chimpanzees, one at about the same rate, and one even faster. So although the averages may be fairly similar, it's an oversimplification to equate each australopithecine's development to that of the great apes. Until scientists can study Taung using synchrotron imaging, we'll continue to assume that it died at around 4 years old.

If the australopithecines didn't share our long childhood, might the first members of our own genus have done so? Some scholars believe that hominin life history began to shift toward a longer development period with the origin of *Homo erectus*, or shortly thereafter.[21] The best-known *Homo erectus* juvenile is an exceptional male skeleton from a Kenyan site near Lake Turkana, dubbed "Turkana Boy." Conventional estimates of his age range from 7–15 years, with the younger figure based on chimpanzee development models and the older on modern human reference samples.[22] Yet, as we've seen with the australopithecines, fossil hominins didn't grow up exactly like living species.[23] For example, Turkana Boy's teeth and bones grew at different rates, which differ from those of humans living today. The estimate of his age at death from tooth growth lines, 7.6–8.8 years, is more similar to an ape-like pattern. It's unlikely that he shared an elongated period of dental or skeletal growth with modern humans, but more comprehensive studies of long-period lines inside his teeth are needed to make precise comparisons with australopithecines or living apes.

One of the hottest topics in this field is whether Neanderthals and modern humans differed in their life history.[24] Scientists have explored many types of evidence to resolve this question, including patterns of tooth wear, developmental stress, and tooth calcification. In chapter 1, we learned how teeth are typically physically sectioned into thin slices in order to view their microstructure. This destructive approach has limited the study of most hominin fossils—including the Taung Child and the Turkana Boy—although in a few instances curators have consented to allow sectioning of Neanderthal teeth. These opportunities have provided important information on crown formation times, age at death, and elemental chemistry, although sample sizes have been too small for rigorous comparison with living humans.

Fortunately, a nondestructive approach has been developed using the "super microscope" at the European Synchrotron Radiation Facility in Grenoble, France. In the beginning of this book, I discussed my rewarding academic partnership with Paul Tafforeau, who was instrumental in developing synchrotron imaging for the study of tooth growth in fossils. Once we verified that this technique provides information comparable to the thin sections I prepare in my lab, but without needing to open teeth through cutting, Paul and I decided to investigate the unresolved issue of whether Neanderthals share a developmental pattern with modern humans.[25] We spent several years working with curators and other paleoanthropologists to collect information on as many juveniles as possible. The first results came from a fossil that is now believed to be one of the earliest members of *Homo sapiens*—a 300,000-year-old juvenile from Morocco (figure 6.3).[26] We scanned a tiny chip of enamel from its incisor, which provided the crucial information needed to determine that the youngster died at 7.8 years of age. Importantly, the teeth in its jawbone were developmentally similar to living human children of the same age, meaning that our growth pattern dates back to at least 300,000 years ago. At the time, I was surprised by the amount of press coverage our work elicited, since I felt that we needed to sample other fossils to fully understand the significance of this finding.

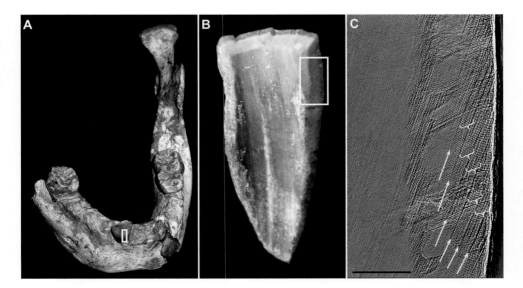

Figure 6.3

Jawbone and enamel from a 300,000-year-old fossil *Homo sapiens* child. (A) The mandible showing the location of the incisor tooth enamel (white box) imaged with the synchrotron. (B) Enamel fragment with the area of interest (on right) shown in the white box. (C) Synchrotron image showing long-period growth lines (white arrows) with 10 daily lines between them (white brackets). The scale bar is 0.2 millimeters. Reproduced from Tanya M. Smith et al., "Earliest Evidence of Modern Human Life History in North African Early *Homo sapiens*," *Proceedings of the National Academies of Science USA* 104 (2007): 6128–6133.

Paul and I then turned our attention to several Neanderthal fossils, with whom humans shared a common ancestor more than 500,000 years ago. We wanted to clarify whether the slow development of the 7.8-year-old *Homo sapiens* individual is only found in our species, or is part of a more ancient pattern. Bromage and Dean had established that australopithecines and early *Homo* were not like us developmentally, so if Neanderthals didn't show our prolonged growth period, it was unlikely to be found in our common ancestor with them—making humans unique in this regard. Numerous curators carried their precious cargo to Grenoble and waited patiently while we bombarded fossil teeth and jaws with lethal X-ray doses in a lead-shielded room—collecting thousands of radiographs of each. The resulting synchrotron images allowed us to measure tooth formation and calculate the ages of several important juveniles, including the first fossil hominin

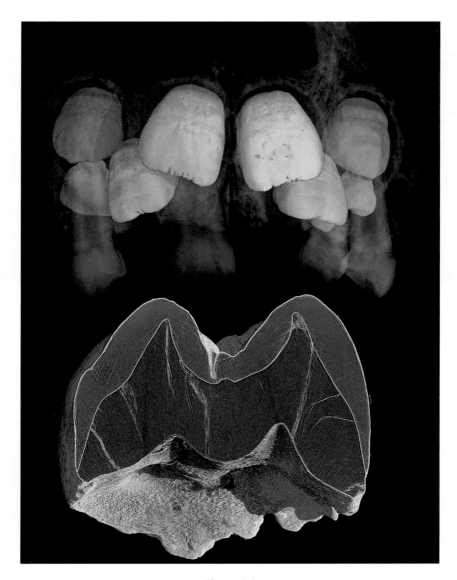

Figure 6.4
Baby Neanderthal upper jawbone showing teeth developing inside the upper jawbone (above), and the permanent molar (below) via synchrotron imaging. Compare to the photograph of this individual in figure 2 in the introduction to this book. Reproduced from Tanya M. Smith et al., "Dental Evidence for Ontogenetic Differences Between Modern Humans and Neanderthals," *Proceedings of the National Academies of Science USA* 107 (2010): 20923–20928.

child to be discovered (introduction, figure 2). The Belgian Neanderthal—discovered with bones from now-extinct rhinoceros and mammoth species—had been thought to be 4–6 years old when it died.[27] Powerful synchrotron X-rays revealed that it actually died at a mere 3 years of age (figure 6.4), by which time it had already grown a remarkably large brain—a point we'll return to in the following section. While the true age might differ by a week or two, Paul and I felt confident in reporting a precise age of 3.0 years for this fossil, since synchrotron imaging revealed beautifully clear growth lines inside the molars, including one formed during the individual's birth!

It turns out that Neanderthal teeth took shape more rapidly than those of humans, including the North African juvenile and one of the earliest fossil *Homo sapiens* individuals to have left Africa 90,000–100,000 years ago. When we compared the ages determined from growth lines to those predicted from human growth standards, we found that the standards overestimated the true ages of Neanderthals. These differences in tooth development persist from early to late childhood, adding to a growing body of evidence that subtle developmental differences exist between humans and our Neanderthal cousins, including the growth of the head and rest of the skeleton.[28] Our team concluded that prolonged dental development appeared after humans' evolutionary divergence from the common ancestor we once shared with Neanderthals. Ongoing research is working to relate differences in tooth growth to life-history characteristics such as weaning age, age at reproductive maturity, and life span. For example, in chapter 7 we'll explore how investigations of tooth chemistry have yielded the first report of weaning in a Neanderthal—an exciting breakthrough built on the faithful time records in teeth.

I Like Big Brains

In the preceding chapter, we identified a number of anatomical features that distinguish humans from other living primates, such as our style of walking on two legs and our big brains. Human brain size is especially impressive when contrasted with the size of our bodies. Let's consider, for a moment, one of our evolutionary cousins. Gorillas are the largest apes living today—eclipsed in the past only by the extinct Asian fossil *Gigantopithecus*. Gorilla males weigh roughly three times more than human males, yet their brains weigh only one-third as much as a human brain! When we survey the entire hominin lineage stretching back 7 million years, brains as large as our own species first appear in the fossil record around 400,000 years ago (figure 6.5). This transition is important for understanding the evolution of human development. Large brains require a lot of energy to grow and function properly, a requirement that may even influence when babies are born.[29] We don't really understand what triggered this heady evolutionary expansion or how it relates to changes in our life history. Some scholars regard brain size as one of the main pacemakers of development, although whether the association between big brains and slow life histories is due to causation or coincidence is up for debate.[30]

Biological anthropologists often invoke a special relationship between teeth and brains.[31] For example, in her landmark 1989 study, Holly Smith found that adult brain size is strongly associated with first molar eruption age across primates. Species with large brains, such as great apes, erupt their teeth at later ages than more modestly endowed monkeys or lemurs. She later went on to use this relationship to estimate when first molars would have erupted in fossil hominins with well-preserved skulls. Because impressions of soft brain tissue are rarely preserved, scientists often measure the space within the braincase, known as cranial capacity, as a proxy for brain size. Yet when Smith and colleagues reanalyzed these data a few years later, it became apparent that evidence from dental development and cranial capacity points to a contradictory picture for early members of the genus *Homo*.[32] Their brains began to get larger in comparison to australopithecines and apes, which is particularly evident by the time *Homo erectus* appears on the scene—yet their tooth growth is nearly indistinguishable from the former groups.

Other scientists have posited that brains and teeth may show similar growth rates that differ from increases in height or weight throughout childhood, as teeth are housed in facial bones that must join the braincase. A variation on this theme is the idea that approximately 95% of human adult brain size is achieved by the time our first molars erupt around age 6.[33] However, this oversimplifies what is known about the development of the head. Skeletal and neural aspects reach maximum size at different ages, including head circumference, cranial capacity, total brain volume, and individual lobe volumes.[34] Daniel Lieberman underscores this point in his tome *The Evolution of the Human Head*, noting that the parts of the skull that surround the brain mature more quickly than the face and the teeth. Furthermore, internal

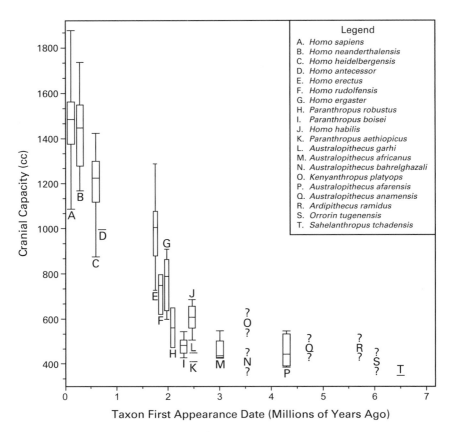

Figure 6.5

Hominin cranial capacity (space inside skull) over 7 million years of evolution. Modified and reproduced with permission from Matthew M. Skinner and Bernard Wood, "The Evolution of Modern Human Life History. A Paleontological Perspective," in *The Evolution of Human Life History,* eds. Kristen Hawkes and Robert R. Paine (Santa Fe, NM: School of American Research Press, 2006).

rewiring occurs in our brains well into the third decade of life—maturing long after our teeth have finished growing—and this fine-tuning is difficult to capture in estimates based on cranial capacity.[35]

Could brain or body size be a better predictor of life history than first molar eruption age or other aspects of dental development? It would be convenient if this turns out to be true, since we know more about the brains and bodies of fossil hominins than about how long it took them to erupt their molars. The idea here is that the bigger the brain or body, the slower the pace of life history.[36] Yet these statistical associations tend to be based on broad analyses of different types of primates. When we zero in on great apes and humans, the relationships don't hold up so well. For example, gorillas have bigger brains and bodies than orangutans, but faster life histories—which is the opposite of what would be expected from the broader primate pattern. Humans fall into an odd place in the ranking; we've already discussed how our bodies are smaller than gorillas, yet our brains are larger. These exceptions make me hesitant about using either to predict aspects of life history in fossil hominins. This is reminiscent of the problem with using first molar eruption age to predict weaning age, which doesn't work terribly well in great apes or humans either.

Despite these limitations, documenting tooth development remains important for tracking brain growth in fossil hominins. This is because large brains can come about through fast growth, long growth periods, or both—pathways that vary across primates and are not fully explained by evolutionary relationships.[37] We've seen how microscopic tooth growth can be used to determine the ages of ancient children—supporting unbiased comparisons with living apes and humans. This becomes particularly important when asking whether differences in rate or timing exist between closely related species, such as humans and Neanderthals, whose brains are more similar to one another than to apes or early hominins.[38] For example, the 3-year-old Belgian Neanderthal in our study of dental development has an estimated cranial capacity of 1,400 cubic centimeters—larger than most human adults! Reaching this size by age 3 required faster growth rates, and it was likely on track to complete its brain growth earlier than humans as well. This kind of evidence adds more weight to the hypothesis that a long, slow developmental pattern is unique to our own species.

How Old Is Old Age?

Our final exploration of life-history evolution makes use of wear patterns on teeth. Recall that marked tooth wear was very common prior to industrial food production. In chapter 2, I explained that Loma Miles was able to estimate how much a first or second molar typically wears down over 6 years—the time between when it erupted and when the next molar erupts. He extended this information to predict adult ages from different stages of wear of the third molar. Non-industrial people who died with lightly worn third molars were likely to be in their 20s or 30s, while those who had ground away the surfaces of their third molars probably survived into their 40s and 50s.

A clever study of the fossil record built on this idea to distinguish hominins who died in young adulthood from those that reached an older age.[39] Rachel Caspari and Sang-Hee Lee examined 768 individuals' molars—an impressive number given the limits of the fossil record—and found that relatively few australopithecines survived to old age, while slightly more early *Homo* and Neanderthal individuals did. Interestingly, many more fossil *Homo sapiens* individuals had heavily worn third molars. It appears that there was an increase in elderly individuals after our species diverged from the ancestor shared with Neanderthals. This is even more impressive when we consider that they had earlier erupting molars than *Homo sapiens*, along with thinner enamel—conditions that made it more likely that Neanderthals would "look old" sooner. The fact that prehistoric members of our species survived long enough to wear down their thick-enameled third molars appears to be part of an overall lengthening of our life span, and possibly, our life history.

Anthropologists distinguish life span, the maximum time any individual lives, from life expectancy—the average age at death in a population. Modern humans can live longer than other primates (table 6.1), and apparently longer than ancient hominin species, but a study of recent human populations concluded that large cohorts of senior citizens are a new phenomenon.[40] It turns out that the life expectancy of hunter-gatherers and non-industrial populations is less than half that of humans from industrialized nations. Historic trends from Sweden illustrate this within a fairly close-knit, homogenous population. Between 1751 and 1759, the average Swede was expected to live until about 40 years of age. Their life expectancy has progressively increased over time, exceeding 80 years by the year 2000. This gap in life expectancy between industrialized and non-industrialized human populations is actually greater than the difference between non-industrialized humans and nonhuman primates. The authors point to recent social, economic, and public health advances that have shifted our biology more rapidly than millions of years of evolution. We'll come back to this point at the end of this book when we consider whether humans are still evolving.

Evidence for the lengthening of human life expectancy complements Caspari and Lee's examination of the fossil record; life span may have increased with the origin of our species, while life expectancy has rapidly jumped over the past few centuries. This is important because studies of tooth wear are a rather coarse way of getting at differences in life history, particularly for juvenile primates.[41] For example, some anthropologists have used patterns of tooth wear in great apes to predict when infants stop relying on mother's milk. Unfortunately, I've not been able to validate this in my research, as visible patterns of tooth wear in chimpanzees do not align with when they start eating solid foods or cease suckling. Similarly, comparisons of tooth wear in young Neanderthals and fossil *Homo sapiens* have been used to suggest that our species began supplementing mother's milk with solid foods at earlier ages than Neanderthals. Yet the ages in this study for the introduction of solid food are unlikely to be correct, since they are much later than those of modern hunter-gatherers or any nonhuman primate, including the late-weaning orangutans. Additional study is needed to disentangle such variables as enamel thickness, dietary variation, and eruption schedules prior to comparing tooth wear across different primates. In the following chapter, we'll turn our attention to how dietary grit and food processing methods also affect the rate and degree of tooth wear, as well as interpretations of what ancient hominins were eating.

Human life history is best understood from an evolutionary perspective, meaning that heritable characteristics supporting the birth and survival of healthy offspring will increase in a population over time. In the conclusion of this book, we'll see that our development isn't static; it has continued to change over the past few centuries. You may be wondering why human life history differs from other primates. Numerous theories have been proposed to explain this, although many of these are difficult to test due to the imperfect nature of the fossil record.[42] Some biological anthropologists believe that our extended childhood and life span resulted from a dietary shift to large, nutrient-dense food. This is based on the idea that hominins began to rely on diets that required sophisticated behaviors, such as hunting dangerous game or gathering ephemeral seasonal foods. Learning these skills may have required a long period of development, during which adolescents could take advantage of intergenerational transfer of knowledge—not unlike our contemporary educational system.

Other scholars propose that our long childhood is a consequence of having large, metabolically demanding brains. The human brain's appetite for energy peaks around age five—a few years before it stops growing in size.[43]

This is also the age when body growth slows to a proverbial crawl, possibly due to the excessive energy demands of our brains, which actively rewire as they increase in size. As the energy needs of the brain abate, body growth speeds up—culminating in an adolescent growth spurt. The growth period in our teen years is often heralded as another unique human trait, since there is no evidence of this pattern in hominins predating our own species.[44] Whatever the ultimate cause of our long developmental period, it has accompanied the establishment of diversified diets, complex social systems, and sophisticated cultural practices. In the section that follows, we'll see how these have left indelible impressions on our teeth.

III   Behavior

Many skeletal collections exhibit teeth that during life had been called on to do unusual things.

—Albert A. Dahlberg, DDS (quote from p. 172 of "Analysis of the American Indian Dentition," in *Dental Anthropology*, ed. Don R. Brothwell (New York: Pergamon Press, 1963): 149–177)

7 Paleo Dining

"Paleo" is all the rage these days. This diet and lifestyle trend promises to improve health and well-being in our increasingly complex world by espousing a return to the habits of our ancestral past. While encouraging to those of us who professionally ponder human biology and behavior from an evolutionary lens—these cultural currents run the risk of oversimplifying our prehistory. Loren Cordain asserts in *The Paleo Diet* that he has "been able to determine the dietary practices of our hunter-gatherer ancestors" and that until 10,000 years ago "everyone on the planet ate lean meats, fresh fruits, and vegetables."[1] This is a bit of a stretch. Please don't think that I'm suggesting the Paleo Diet isn't a healthy way to eat; that's a question for dietitians and nutrition scientists to tackle. What I take issue with, after weighing the evidence detailed in the following pages, is the idea that there was *one* diet consumed by our ancestors, that we can itemize it, and that it remained constant prior to the development of agriculture.

Cordain and his Paleo Diet fans draw upon evidence from the fossil record and observations of modern hunter-gatherers to infer the behavior of earlier hominins.[2] For example, he points to signs of animal butchery around 2.5 million years ago, concluding that a novel diet with significant protein catalyzed the development of large brains in the genus *Homo*. This isn't a new idea; anthropologists have proposed and debated similar ideas about the evolution of the human diet for decades, including the influential "Man the Hunter" model of human nature from the 1960s and '70s. Yet Cordain overlooks alternative perspectives about the significance of meat in the diet of early *Homo*, and whether individuals routinely obtained lean-meat from freshly hunted kills, or scavenged bone marrow and brain tissue from abandoned carcasses.[3] We now know that australopithecines, the group of hominins that preceded our own genus, began making tools and cutting animal bones more than 3 million years ago. Thus some form of meat consumption occurred prior to the evolution of our genus—millions of years before big brains evolved—which casts doubt on the primacy of meat eating in establishing our humanity. Like most scientific disciplines, biological anthropology has a rich history of refining or replacing ideas that were persuasive when they were first proposed.

Exploring the evolution of the human diet over the past seven million years is an ambitious undertaking. Peter Ungar and colleagues tackled this in a book that highlighted the "known, unknown, and unknowable."[4] While the look and feel of our choppers may be an obvious source of information about the human diet—the crests and cusps we put into service every few hours noshing snacks or meals are not the most powerful evidence we can draw from. Paleoanthropologists collect clues about ancient diets from the microscopic structure and chemistry of hominin dental remains, as well as the discarded animal bones and stone tools found with them. We also work closely with primatologists, who study the diet and feeding behavior of our nonhuman kin, and consider insights into ancient environments from paleoecologists and climate scientists. Here we'll dig into the study of past diets using the fossilized teeth of ancient species. New methods and promising analytical tools continue to enrich our understanding of this essential behavior. As Ungar demonstrates, there are more than enough opinions to fill an entire book. For those of you who aren't on the Paleo Diet, grab your popcorn and read on!

Show Me Your Teeth and I'll Guess What You Ate

Teeth are invaluable for obtaining nutrients and energy to grow up, keep in good health, and pass our genes on to the next generation. An essential body part like our dentition has been under strong evolutionary pressure to work efficiently. Until recently, a developmental mismatch or serious malfunction could lead to starvation and premature death. Comparative anatomists, including Aristotle and George Cuvier, have highlighted an adaptive relationship between tooth form and function—meaning that the shapes of teeth reflect their particular use. We discussed this in chapter 4 when considering the origin of mammals. Their dentitions have a streamlined interlocking design, with specialized incisors and canines at the ready for taking food in, and multi-cusped premolars and molars for breaking it down (figure 4.4).

Primates have built upon this—particularly those that specialize on a narrow range of foods. Ironically enough for those of us concerned with human uniqueness, the shape of the human dentition is pretty simple compared to other primates and mammals. We're often categorized as dietary generalists, or omnivores, as our teeth and guts are not tuned for a narrow plant-based diet like many herbivores (think flat-toothed cows) or a meat-based one relished by carnivores (think pointy-toothed lions). When comparing primates as a whole, basic dietary divisions are apparent between those whose teeth have tall cusps that serve well for eating leaves, and those with low-cusped teeth for eating fruit and seeds. Unfortunately for us, these distinctions are less helpful when interpreting the dental remains of fossil hominins.

In primates with highly varied diets, including the great apes, tooth shape doesn't always reflect food preferences. Although they relish eating ripe fruit when it is available, great apes are forced to eat other foods when fruit is scarce. Ecologists refer to these choices as fallback foods—essential staples that help primates survive the lean months of the year. For gorillas, this may mean eating tough vegetation, leading to the evolution of tallish molar cusps with long ridges for shearing pithy stems and leaves. Orangutans rely on different foods during times of fruit scarcity, such as tough bark or hard seeds. Their low-crowned molars have fine ridges to clamp objects when applying considerable force (figure 7.1).

Ungar suggests that subtle differences in the shapes of hominin teeth also reflect varied consumption of fallback foods.[5] Yet reconstructing diets from tooth shape is complicated—all fruits, seeds, and leaves are not created equal. For example, tender new leaves produced seasonally will toughen up as they mature. Imagine biting into a ripe peach and then the outer shell of a coconut—these two fruits will respond very differently to the same amount of force during biting. Furthermore, primates with diverse diets have evolved dentitions that reflect a compromise among foods with varying degrees of hardness and toughness. The simple relationships between tooth form and diet may be blurred by competing demands, which might be the point where Cuvier's maxim "show me your teeth and I will tell you who you are" breaks down. It's not possible to simply look at the teeth of fossil hominins and know which specific foods they ate.

Another exploration of prehistoric diets involves studying enamel thickness. In chapter 5, we discussed the fact that the tooth crowns of humans and most fossil hominins are capped by a substantial amount of enamel,

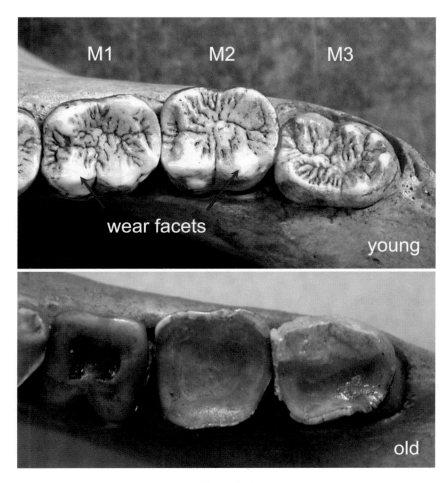

Figure 7.1
Molar macrowear in a young (above) and old (below) orangutan dentition. Orangutan
jaws courtesy of the State Anthropological Collection (Munich, Germany).

which exceeds that of our African ape cousins (figure 5.3).[6] Following
Cuvier's logic of predicting function from form, a number of anthropologists have speculated about what such thick enamel was good for. In 1970,
primatologist Clifford Jolly proposed an influential idea that drew on studies of baboons, ground-dwelling African monkeys with thick tooth enamel
who frequently consume grass seeds. Jolly suggested that thick enamel in
baboons and hominins may be adaptive for breaking up small, hard, round
objects through crushing and rolling, as occurs in milling machines. In this

case, thick enamel is advantageous for resisting wear caused by abrasive particles in or on the food items. This led to the "seed-eater hypothesis"—particularly germane for the australopithecines that once lived in environments similar to those of modern African baboons. Another influential idea took root a decade later, when Richard Kay demonstrated that living primate species that ate hard fruit had the thickest enamel—leading to a competing "nut-cracker hypothesis" for hominin dietary evolution. In this scenario thick enamel kept teeth from cracking when forcefully chewing hard foods.

So which is it? Did our ancestors eat lots of small abrasive food items that needed to be finely ground, or did they dine on larger hard foods that required some muscle to masticate? To date, it's not clear which model best explains the evolution of thick enamel. These ideas aren't mutually exclusive—thick enamel could have served to resist abrasion when eating small, hard foods *and* to allow hominins to bite down on large, hard foods with a degree of force that may cause weaker teeth to fracture. If this isn't complicated enough—the plot thickens with additional suggestions that enamel thickness reflects the consumption of fallback foods rather than common or preferred dietary items. My PhD advisor, Lawrence Martin, and his PhD advisor, Peter Andrews, proposed that thick enamel was an adaption for a more diverse diet.[7] This may have been driven by colonization of more temperate environments during the course of our evolutionary history in Africa. In strongly seasonal environments—especially those away from the equator—living primates must rely on tougher or harder foods during lean seasons than those in more consistent tropical ecosystems.

I decided to test this idea in macaques—Old World monkeys found in disparate environments, including temperate Japan, tropical Southeast Asia, and seasonal cedar forests in the mountains of Morocco.[8] My team measured enamel thickness in several hundred molar teeth, discovering that macaque species in seasonal environments had thicker molar enamel than those in the tropics. This supports Martin and Andrews' idea that thick enamel may be especially beneficial for hominins that lived in temperate seasonal environments, who likely consumed tough or hard foods from time to time. But before we go too far with this explanation, I'll disclose that macaques have highly variable diets in all the environments they inhabit. While thickened enamel and seasonal environments appear to relate in macaques, it's not possible to itemize a species' diet based on the amount of enamel on a molar tooth. It appears that we've reached a similar conclusion as the aforementioned studies of primate tooth shape. Let's turn to other approaches to uncover how our diet has evolved since we first began to forage on two legs.

Tooth Wear: The Last Supper?

In an ideal world, our understanding of past diets would be built on direct evidence of eating behavior, rather than extrapolations from tooth shapes or enamel thickness. One seemingly obvious example is the wear on an individual's dentition. In chapter 2, we learned how tooth wear has been used to estimate the age of skeletal remains. This irreversible process begins when food or opposing teeth rub against tooth surfaces, which create a shiny polished look. Add some abrasive particles during forceful chewing, and tiny microscopic scratches and pits form. Wear facets are produced when teeth repeatedly come into contact with each other, flattening the enamel. These facets grow larger with continuous use, eventually revealing the underlying dentine, while blunting cusps and crests. Over time, islands of exposed dentine on the chewing surface merge, ultimately leaving a plateau of soft dentine and pulp (figure 7.1). At this advanced stage of wear, tooth infection and loss are common, causing great discomfort and hindering chewing efficiency.

Few of us will ever reach this stage naturally, with modern hygienic diets, clinical care, and myriad food processing methods intervening. A casual inspection would probably convince you that fossil hominins wore through their teeth more often than humans living today. Comparisons of visible wear patterns show some differences within our species.[9] For example, the teeth of ancient hunter-gatherers are heavily worn compared to those of early agriculturalists, who likely ate a softer, less fibrous diet than people relying solely on wild foods. While overall wear patterns may hint at the basic dietary practices of these two groups, they aren't particularly helpful for identifying the diets of hominins that predate our own species. This is because all foods were gathered, scavenged, or hunted prior to the invention of agriculture 10,000–15,000 years ago.

Given the limitations of visible tooth wear patterns for answering questions about our deeper past, considerable effort has gone into finding other ways to decode clues ground into teeth during biting and chewing.[10] One influential approach documents the microscopic pits and scratches found on wear facets. Initially scientists set out to test whether this *microwear* is simply caused by tooth-tooth contact or whether it is due to chewing food. The teeth of stillborn guinea pigs from a research facility were compared with those from adult guinea pigs. While in the uterus, young guinea pigs grind their teeth against each other, but it turns out that these do not show any microscopic scratches or pits. In contrast, laboratory-fed and wild adult guinea pigs have clear signs of microwear. Teeth can wear down by contacting one another—as any habitual tooth grinder can unfortunately attest—but microscopic pits and scratches appear to be caused by something that has been eaten.

Anthropologists may somewhat irreverently refer to microwear as a "last supper" signature, which isn't really the case for living humans or laboratory animals fed soft diets, who wear their teeth more slowly than prehistoric people or wild animals.[11] These tiny marks are typically studied from plastic replicas of fossils using powerful microscopes. In order to interpret microwear patterns on ancient teeth, it helps to start with modern individuals whose diets are well understood. A study of African hyraxes, small ground-dwelling mammals with plant-based diets, is an informative example.[12] The authors collected skeletal remains from two species that live in the same region, and found extensive linear striations on the species that mainly ate grasses during the wet season. These were not found on the other hyrax species, which largely fed on bushes and trees during this season. Moreover, their wear patterns varied according to seasons, as both hyrax species showed less marks during the dry season, when they both relied on bushes and trees. Thus microwear is overwritten by next season's diet.

This study led to an influential hypothesis that these patterns can separate mammals who eat grassy ground cover from those who eat shrubs and trees. Similarly, leaf-eating howler monkeys are known to have more striated microwear than nearby capuchin monkeys, whose highly pitted microwear appears to be caused by chewing hard food such as seeds.[13] This basic dichotomy of scratches versus pits, and the complexity of tooth surface relief, is often used to infer the properties of fossil hominin diets (figure 7.2).[14]

A classic example is *Paranthropus robustus*, a South African hominin with massive teeth and jaws, which shows excessive pitting on tooth surfaces. Microwear specialists interpret this to mean that it consumed hard food at least occasionally, similar to the seed-eating capuchin monkeys. Oddly, a closely related species found in East Africa, *Paranthropus boisei*, has a nondescript microwear pattern that implies it ate neither hard nor tough foods. Yet both species have similar facial structure and large, thick-enameled

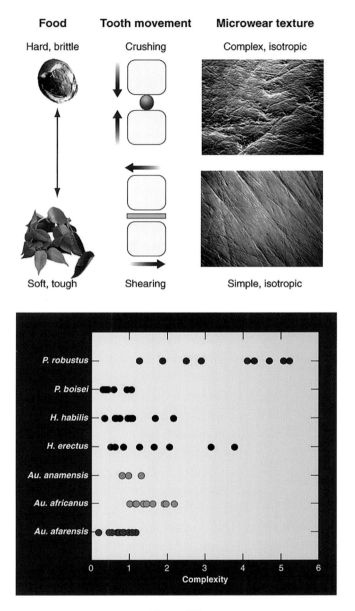

Figure 7.2
Illustration of microwear patterns and their distribution in
fossil hominins. From Peter S. Ungar and Matt Sponheimer,
"The Diets of Early Hominins," *Science* 334, no. 6053 (Oct 2011):
190–193. Reprinted with permission from the American
Association for the Advancement of Science.

teeth—which seem especially well adapted for crushing and grinding hard objects.[15] We'll return to this discrepancy below when we consider evidence from tooth chemistry. Microwear on African *Homo erectus* teeth indicates a varied diet, with tougher or more brittle foods than other early *Homo* specimens. Again this result seems discordant with the more delicate facial architecture of *Homo erectus* relative to most australopithecines—a group historically regarded as having faces and jaws built to tuck into tough or brittle foods. Most *Australopithecus* species that lived around 2–4 million years ago show less pitting than *Paranthropus robustus* and *Homo erectus*, and fewer scratches than leaf-eating primates, which has been interpreted to mean that they were neither routinely eating hard objects nor shearing tough foods such as leaves. It seems we cannot determine what australopithecines were eating from the microwear evidence alone.

Tooth microwear varies depending on the types of foods available during the season when the individual died, as we learned from the comparison of African hyraxes. Modern hunters-gatherers consume seasonal foods of differing degrees of hardness and toughness, which was likely the case for the Neanderthals that managed to survive several glacial periods in Northern Eurasia. Studies of their tooth wear report they relied on both plants and animals, depending on where they lived.[16] Neanderthals from forested regions in the Mediterranean and the Middle East show a stronger signature of plant consumption than those from Northern Europe. Overall wear profiles in Neanderthals align them with recent hunter-gatherers who had meat-heavy diets.

Yet meat does not create its own microwear signature. Scientists assume that traditional methods of food preparation introduced grit or other abrasives that produce microwear during chewing.[17] Stone tools used to butcher carcasses might be one source of abrasive material. In an experimental study, a human volunteer consumed cornmeal muffins prepared from corn ground with a traditional Native American sandstone tool, eating them daily for a week.[18] This led to a massive increase in microwear when compared to his typical American diet. Eating cornmeal muffins produced with a finer-grained stone tool also produced microwear, although not as much as from the sandstone tool. Thus abrasives may explain the pronounced wear of many prehistoric Native Americans who relied on ground corn. Ancient Aboriginal people also have extreme tooth wear, possibly due to the use of large flat stones to mill wild seeds in certain parts of Australia.[19]

The issue of grit has long plagued microwear studies, as it is often difficult to tease apart the signature of environmental grit consumed unintentionally from plant-based foods with abrasive properties.[20] Anyone who's spent time at the beach has become intimately familiar with silica, the key component of sand that makes it gritty. Silica occurs naturally as quartz, a common mineral prized by rock collectors and jewelers alike for its striking crystalline forms. Plants and especially grasses incorporate microscopic silica particles as a structural defense against herbivores. These *phytoliths*, or "plant stones," produce highly striated microwear that speeds up tooth wear when they're consumed, serving as a potential deterrent for herbivorous animals. Experimental work has shown that environmental grit can create a microwear pattern that can be mistaken for grass consumption. For example, a laboratory colony of opossums fed cat food, plant fiber, and chitin showed similar patterns of moderate scratches, while one individual that had grit added to its diet showed heavy microwear similar to the wild grass-eating hyraxes discussed earlier. Thus, true dietary signals in tooth microwear may be obscured by the inadvertent consumption of grit, which may be common in open savannah-like environments as well as in forested regions.[21]

If this wasn't discouraging enough, even more factors can influence the formation of microwear.[22] For example, enamel that accidentally chips off can indent the crown—and chipping is common in mammals that feed on hard objects, as well as early hominins. Moreover, the angle of contact between teeth during chewing also impacts the formation of pits and scratches. One experiment demonstrated that pits were formed when teeth contacted food during perpendicular biting, scratches were formed where teeth slid past one another during a parallel approach, and a mix of pits and scratches was formed when teeth met at a 45-degree angle. Importantly, the same type of food was consumed during each trial. This finding seriously complicates dietary reconstructions of fossil hominins. At this point, it seems that our search for an accurate understanding of ancient diets must continue.

Tooth Chemistry: Out of the Mouths of Babes

Anthropologists often repeat the trope "you are what you eat," since it points to an important truth about our bones and teeth. The elemental building blocks of skeletal formation, such as calcium, oxygen, carbon, and nitrogen, come directly from what we eat and drink. Before birth and while nursing, our mothers provide these important elements, along with tiny amounts of nonessential metals, which are permanently recorded by our mineralizing teeth. Once we begin to eat solid foods, the chemistry of teeth reflects this change. When the third molars finish forming around 20 years old, they stop adding dietary elements as well as the tiny growth lines that are near and dear to my heart.[23] However, these youthful chemical records are retained throughout life unless they are worn away. In contrast, our bones replace early elemental signatures during routine lifelong remodeling and repair, meaning that an adult's bones present a different dietary history than their teeth.

As you might imagine, scientists eagerly capitalize on this system to reconstruct ancient species' diets. Tiny amounts of enamel and dentine are first removed with drills, lasers, and ion beams, which are then fed into large machines called mass spectrometers that characterize atoms, molecules, and ions in the sample. This approach also allows us to explore nursing behavior in humans and other primates.[24] My colleagues and I have spent the last few years studying a metal called barium in thin sections of teeth, which gives us special insight into the nursing process. In the previous chapter I explained why weaning is a key life-history signal; juveniles become nutritionally independent when they cease nursing, allowing their mothers to muster energy for their next offspring. Before I tell you what we've found, I need to explain why this element is so useful. Barium comes from the maternal skeleton along with calcium—a concentrate in mothers' milk to help infants build their own bones and teeth. It isn't an essential nutrient like calcium, and in fact excessive amounts can be toxic. Because infant guts are highly receptive to calcium, and barium is chemically similar to calcium, both elements are readily absorbed into the bloodstream from milk and ultimately written into the hard tissues of each nursing infant.

You may have heard of barium, since it is used for clinical imaging of the digestive tract. This metal absorbs X-rays and helps reveal the structure of soft internal tissues that are otherwise hard to see with radiographic imaging. While barium is naturally found in foods besides mother's milk, adult digestive tracts aren't very good at absorbing it. A few years ago I had to drink a liter of artificially sweetened barium sulfate in preparation for a CT scan, and was comforted to know that my body wouldn't be retaining much of this unforgettable drink!

Returning to the dietary record in teeth, my collaborators Manish Arora, Christine Austin, and I have shown that barium concentrations change after the neonatal (birth) line is formed, which is due to the onset of nursing. Barium values before birth are very low, since the placenta shields the developing fetus from maternal stores—most of which are locked inside mom's bones and teeth. Once an infant begins to suckle calcium- and barium-rich milk, the values in teeth increase. These values increase further when human infants are given commercial infant formula, which contains even more barium than human milk. Our team also studied the teeth of several macaque monkeys from a laboratory colony. Their barium levels rose after birth—just like human infants—peaking during exclusive nursing, and declining when the monkeys started eating solid food. One of the youngsters had been separated from its mother and hospitalized for several weeks due to illness. This led to a premature end to nursing, and its first molar tooth showed a corresponding steep decrease in barium at this age—confirming the power of this approach to document weaning in juvenile primates. We have since elucidated nursing behavior and weaning ages in several wild orangutans as well.[25] These critically endangered apes may continue to nurse for more than 8 years—longer than any other wild animal—making them especially vulnerable to environmental destruction.

The fact that we can detect and age these early life diet changes is also interesting to public health scientists, as a growing number of studies of humans have found significant health benefits from prolonged breastfeeding.[26] Milk sustains infant growth and immunological development, while providing signals about the environment and the mothers' own wellbeing. Anthropologists have long endeavored to identify whether nursing behavior changed over the course of human evolution, but precise techniques such as these have been elusive until recently. In chapter 6 we learned that—despite initially promising results—first molar eruption ages aren't effective for predicting weaning ages in great apes and humans, limiting their usefulness for probing the hominin fossil record.

Our team applied this elemental approach to document nursing in an 8-year-old Neanderthal, as I had sectioned a first molar previously to calculate its age at death (figure 7.3).[27] I used the method described in chapter 2 to learn that the tooth enamel had started forming 13 days prior to birth, continuing until the individual reached 2.35 years of age. A few years later, Manish and Christine ran their state-of-the-art laser slowly across the section to sample elements in tiny adjacent spots. When I received the barium results, I compared it to my developmental map with suspicion, as we

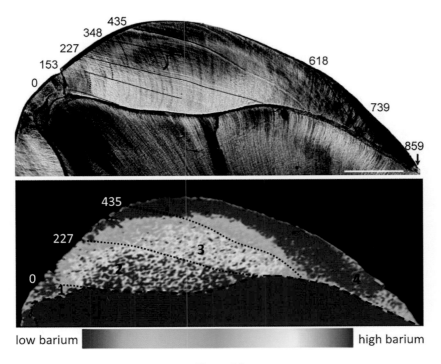

Figure 7.3

Neanderthal nursing behavior revealed by the distribution of barium in a first
molar. Above, age (in days) of accentuated lines (solid blue lines) in enamel
determined from daily growth increments. Below, barium distribution shows four
distinct regions: (1) prenatal; (2) exclusive mothers' milk; (3) transitional period;
and (4) post-weaning. Reprinted from Austin et al., "Barium Distributions in Teeth
Reveal Early-Life Dietary Transitions in Primates," *Nature* 498 (2013): 216–219,
with permission from Nature Publishing Group.

figured that the burial environment would have overprinted the original
dietary information. Yet there were clear zones of barium that paralleled its
enamel development, an unlikely pattern if this was due to contamination
after death. Barium increased at birth, as expected, then dropped after 7
months of age—coincident with a minor stress line. We reasoned that this
transition was due to the introduction of solid food. This is similar to nursing practices in traditional human societies, as mothers tend to introduce
solid foods to infants around 6 months of age. Chimpanzee infants are also
suckled exclusively until this age, adding more support for our interpretation of the Neanderthal child.

As I continued my interpretation of barium in the first molar, I found a real surprise. The elemental signal plummeted a mere 7 months after the infant began to consume solid foods, returning to prenatal levels shortly after it turned a year old. Finding a Neanderthal child that was weaned around 14 months is quite different from both humans and chimpanzees. Children in traditional human societies are completely weaned at 2.4 years, on average, although there is considerable variation among human mothers. Chimpanzees wean their offspring even later, between about 4–7 years old. When I looked more closely at the barium decrease in the Neanderthal, I was reminded of the sick monkey infant who was prematurely separated from its mother in our earlier research. The barium patterns were remarkably similar—a precipitous drop occurred in both individuals, which was quite different from the gradual decrease in macaques that ceased suckling naturally. Another clue was the presence of a very clear accentuated line in the enamel at the same age as the sharp barium decrease, which indicates that this individual also experienced some kind of developmental stress. While we'll never know for sure, I suspect that this young Neanderthal may have been prematurely separated from its mother, resulting in its abrupt and unnatural weaning at 1.2 years of age. By painstakingly combining information from tooth microstructure and elemental chemistry, we're now working to determine when Neanderthals in more favorable circumstances were weaned.

Tooth Chemistry Redux: You Really Are What You Eat

Barium isn't the only element that is useful for learning about an individual's diet (figure 7.4). Perhaps you'll recall from a high school or college chemistry class that elements often exist in multiple atomic states that differ in mass. A classic example is carbon, found in variants called *isotopes* that have mass numbers of 12, 13, and 14—indicated as ^{12}C, ^{13}C, and ^{14}C and read as "carbon twelve, carbon thirteen, and carbon fourteen," respectively. The first two carbon isotopes are stable under natural conditions, while ^{14}C is a radioactive isotope that changes its atomic structure at a constant rate once an organism dies. This process of radioactive decay forms the basis of ^{14}C dating: the less ^{14}C remaining, the older something is. We'll return to the science of dating fossils in the following chapter. More relevant for our exploration of ancient diets is the fact that another carbon isotope, ^{13}C, relates to the foods we eat and can be measured in tooth enamel and dentine.

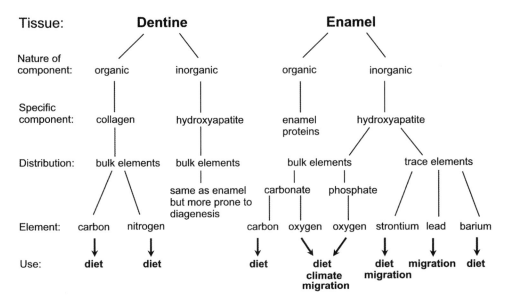

Figure 7.4

Potential chemical analyses of teeth. Hydroxyapatite is the main mineral in dentine
and enamel. Bulk elements are the most common elements in teeth, while trace
elements are more rare. During burial they can be altered by the environment in
a process known as diagenesis, particularly in the porous dentine. Modified from
Tanya M. Smith and Paul Tafforeau, "New Visions of Dental Tissue Research,"
Evolutionary Anthropology 17 (2008): 213–226, with permission from Wiley-Liss.

In the preceding section I detailed how our childhood diets are immortalized in the chemical makeup of developing teeth, and this includes the
solid foods that sustained us once we were weaned off milk. In order to
interpret some signals of sustenance, it is helpful to know that ^{13}C varies
in plants depending on their photosynthetic process. Certain tropical and
subtropical plants—including many grasses, papyrus, and corn (maize)—
capture carbon from the air to produce energy using a water-efficient mechanism known as a C4 pathway. Temperate or cool-adapted plants use an
alternative energy production mechanism called the C3 pathway. These C3
plants are more numerous, making up the great majority of plants in the
world. While their photosynthetic process is less efficient in the tropics,
they are found there too. The key point here is that ^{13}C values are much
greater in C4 plants than in C3 plants, a difference that is also reflected in
the insects and animals that eat these plants, as well as the animals at the

top of the food chain. Thus the teeth and bones of animals that specialize
on C4 grasses, such as antelopes, have higher ^{13}C values than giraffes or
other animals that browse on C3 shrubs and trees. Ancient humans were
also part of this food web, although legitimate questions remain about
when we topped the pyramid as the leading carnivore. Isotopes in teeth are
helping scientists to zero in on the timing of this transition.

Paleoanthropologists have spent the past few decades carefully sampling
carbon isotopes from the teeth of African hominins, and they've found
striking dietary differences spanning several million years (figure 7.5).[28] The
oldest species sampled to date, *Ardipithecus ramidus*, shows the ^{13}C signature
of a predominantly C3 plant diet more similar to living chimpanzees than
to later hominins. A similar result was found for two individuals of *Australopithecus sediba*, despite being separated from *Ardipithecus ramidus* by more
than 2 million years and several thousand miles. In contrast, *Paranthropus*

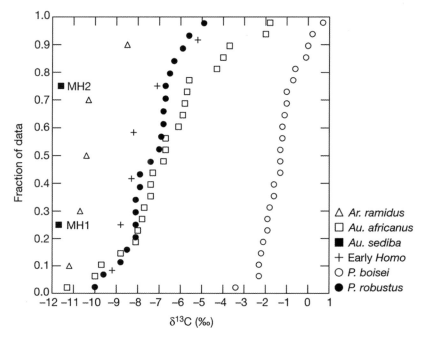

Figure 7.5
Carbon isotope values as a proxy for hominin diets. Each symbol represents
a different tooth sampled from six hominin species. From Amanda G. Henry
et al., "The Diet of *Australopithecus sediba*," *Nature* 487 (July 2012): 90–93, with
permission from Nature Publishing Group.

boisei—the flat-faced robust australopithecine from East Africa—shows an extreme tropical C4 plant signature, implying that it relied heavily on grasses, tubers, and sedges. Scientists can't rule out the possibility, however, that it also ate a considerable amount of insects or animals that consumed C4 plants. Other hominins from this time period—early *Homo*, *Australopithecus africanus*, and *Paranthropus robustus*—show a more mixed diet. The balance between C3 and C4 is variable among members of the Taung Child's species, *Australopithecus africanus*, complicating simple generalizations. Because it isn't possible to distinguish between the consumption of plants and the animals that ate those plants, other approaches are needed to elaborate on these clues and determine when early hominins shifted from a primarily plant-based diet to one that incorporated significant amounts of meat.

Those of you searching for "the" Paleo Diet might be somewhat perplexed. Our early ancestors and their relatives appear to have experimented with a number of different diets as they spread throughout Africa. One of the most striking results is that two robust australopithecine species, *Paranthropus boisei* and *Paranthropus robustus*, consumed very different foods despite having nearly identical teeth and heads. These East and South African species show differences in ^{13}C values and dental microwear, although details of the evidence don't align perfectly.[29] Paleoanthropologists who study their tooth, jaw, and skull form report that something seems amiss, as both of these species were able to bite with considerable force, yet only *Paranthropus robustus* appears to be putting its massive choppers to the test. This debate among scholars of microwear, tooth chemistry, and facial size and shape is an example of how reasonable people can reach different conclusions about our ancient behavior.[30]

What does tooth chemistry tell us about the hominins who spread out of Africa? Unfortunately we have little dietary information for the earliest hominins known to have emigrated. The gap between African and Eurasian dietary studies extends for nearly a million years, partially due to the environmental history of these regions. For example, temperate Europe is almost exclusively home to native C3 plants; thus, comparisons of ^{13}C values aren't useful for looking at the consumption of different types of plants.[31] Scientists who work on European hominins focus instead on a combination of carbon and nitrogen isotopes from dietary proteins, which help to distinguish among carnivores, herbivores, and omnivorous mammals. The catch is that these analyses require organic material—specifically *collagen*, a key protein in dentine and bone. Unfortunately, water, heat, microbes, and chemicals in soil hasten the decomposition of collagen, eventually leading to permanent loss of this powerful dietary evidence.

Individuals less than 100,000 years old from cool climes are the most promising for collagen recovery. Neanderthal bones and teeth make up the majority of study samples, since such stocky hominins were quite at home in chilly Eurasia.[32] They are thought to have obtained a large amount of their dietary protein from hunting large herbivorous mammals. Menu options in the prehistoric past of Europe included filet of mammoth, bison, rhinoceros, and wild horse—most of which have since disappeared. The Neanderthals' isotope values align most closely with mammals they lived alongside, including top-level carnivores such as wolves or hyenas.[33] Modern humans who lived in Europe near the end of the Neanderthals' reign had even higher nitrogen isotope values, suggesting a similarly carnivorous diet that also included foods from freshwater or marine ecosystems.

Were these species on an animal-only diet? Some have noted that it is unlikely any hominin could eat as much meat as carnivorous mammals, since excessive levels of animal protein can be dangerous for humans, especially for pregnant women and infants.[34] One limitation of these studies of tooth chemistry is the fact that protein derived from meat hides the signatures of plants, which are largely invisible in conventional collagen analyses. A new method that targets nitrogen isotopes in amino acids, or individual protein building blocks, may provide finer resolution. The team that pioneered this method reports that Neanderthals may have obtained 20% of their dietary protein from plants.[35] We'll return to this shortly, as these results are more consistent with our final approach to ancient dietary reconstruction.

Dental Plaque as Dietary Flypaper

An exciting new area of anthropological research is the study of tooth calculus, commonly known as tartar. In chapter 3, we learned how this mineralized conglomerate begins as plaque, a sticky biofilm that houses oral bacteria and their metabolic waste products. Bacterial colonies latch on to crowns and exposed roots, giving teeth a "furry feeling" when probed by the tongue. The colonies produce minerals that harden and trap their microbial kin, as well as food particles and cells from the host's mouth. "Fossilized" layers of plaque build up, which can be quite extensive in humans who

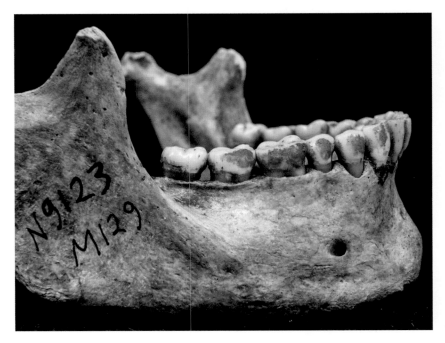

Figure 7.6
Dental calculus (brown stains) on an adult human dentition from the preindustrial
period. Human mandible (968–10–40/N9123.0) courtesy of the Peabody Museum.
Copyright 2018, President and Fellows of Harvard College.

eschew tooth brushing (figure 7.6). Considered the scourge of fastidious
dental hygienists and those of us with gum inflammation, calculus is also
a potent time capsule for investigating the evolution of our diet and oral
health.[36] It sticks around for thousands of years, is fairly impermeable, and
retains its original structure and composition quite well.

Biologists have known that mammalian calculus houses microscopic
remains for over 40 years, but it wasn't until 2011 that Amanda Henry and
her colleagues made front-page news for identifying microscopic plant
remains in Neanderthal tooth calculus.[37] Earlier in this chapter, we discussed how plants incorporate silica, bound up in hard microscopic phytoliths that can wear away enamel, as a structural defense. Scientists can
actually identify plant types from phytoliths and starches trapped inside
the mineralized plaque, although not always with the same specificity as
DNA studies. Starch is a complex carbohydrate that comes in different sizes

and shapes depending on plant type and part. Henry and her colleagues were able to extract a number of these microfossils from three Neanderthals that lived in modern Belgium and Iraq. When her research team compared them to modern plants, they identified date palms, legumes, and grass seeds. Some starches had even been cooked before they were consumed—providing our most specific dietary clue to date!

Cooked starches were also found in the calculus of two Spanish Neanderthals the following year, confirming that this behavior occurred in multiple regions.[38] When starch is heated, it undergoes an irreversible structural change that makes it more digestible, which suggests that Neanderthals intentionally enhanced the palatability of plants through cooking. All living human societies heat food, yet paleoanthropologists have struggled to find secure evidence for the origins of cooking in the fossil record.[39] Neanderthals were able to control fire by the time they appeared in Europe, as the presence of ancient hearths or fireplaces attests. Yet direct evidence of cooking had been previously limited to inferences from burnt animal bones and heated stone tools.

The food we eat contains macronutrients in the form of carbohydrates, fats, and proteins, as well as micronutrient vitamins and minerals. Carbohydrates have been largely invisible in the fossil record—aside from carbon isotopes, which don't distinguish direct consumption of plants from the consumption of animals that eat plants. While isotopic reconstructions once positioned Neanderthals as hyper-carnivores, this has recently been questioned due to evidence from tooth wear and plant microfossils. Neanderthals appear to have ingested grass seeds fairly regularly, including wild relatives of wheat and barley, as well as roots or tubers. Modern humans who inhabited Europe after the Neanderthals also had similar microfossils in their tooth calculus. This evidence flies counter to the argument of the Paleo Diet, which posits that early humans weren't eating cereal grains or starchy vegetables, and thus that these foods should be avoided.

You might wonder how well these plant traces in dental calculus represent an individual's overall diet. Henry and her colleagues wisely tested this using calculus from humans and chimpanzees, both of whose diets they had tracked.[40] They found that microfossils from the teeth of a single human did not capture its diet very well. It seems that this is a sampling problem; in order to get an accurate impression of dietary breadth, many individuals with similar diets must be studied. And since not all plants contain starch or phytoliths, some can't be detected with this approach. Research on chimpanzee dental calculus is consistent with this, recovering phytoliths more commonly than starch—which was underrepresented in comparison to what was actually eaten. Furthermore, both larger starches and phytoliths from the chimpanzee diet were found and identified more easily than smaller plant remains. Henry and colleagues conclude that the microfossil remains in dental calculus may best address questions about the presence versus absence of specific items, rather than revealing the entire scope of plant foods ancient hominins dined on.

Ongoing investigation along these lines has shown that dental calculus traps dietary proteins as well as carbohydrates. Archaeological scientist Christina Warinner and her team have documented the presence of milk whey proteins in the calculus of humans from Europe and Southwest Asia.[41] Samples as old as 3,000 years tested positive for these proteins, and the preservation was so good that her team was able to determine that some proteins derived from cow's milk, while others came from the milk of sheep and goats. Analyses of ancient dietary proteins in calculus are just getting underway, but they appear to hold great potential for tracing the incorporation of plants and animals into the diets of recent humans. Stay tuned for the next dietary fad inspired by these insights!

I'll confess that while reviewing dental evidence for ancient human diets, I've found myself a little confused about which approach to trust, particularly given the methodological caveats and conflicting conclusions. For example, while laboratory and field studies have proven that foods can create pits and scratches on the teeth of living mammals, significant factors complicate the interpretation of hominin diets. Experiments has proven that the same diet can produce varied microwear signatures due to the presence or absence of abrasive grit, alteration in chewing mechanics, and differences in the hardness or toughness of particular foods. Conversely, similar microwear signatures may be produced through combinations of different foods, chewing mechanics, and abrasive contaminants. The most secure way forward appears to be the integration of multiple approaches, bearing in mind the biases or limitations of each. For example, a recent study of *Australopithecus sediba* united evidence from dental microwear, carbon isotopes, plant microfossils from calculus, tooth size, and facial architecture.[42] The current consensus is that this South African hominin had a rather unique diet with a high proportion of C3 plants—possibly including tropical shade- and water-loving plants, as well as fruit, leaves, and wood or bark. The latter items don't sound too appetizing to this hominin!

Similar combinations of methods are shifting paleoanthropological dogma, including the belief that Neanderthal were largely surviving on meat that was hunted or scavenged. Cutting-edge analyses of nitrogen isotopes in Belgian Neanderthals point to a significant amount of plant material in their diets, which aligns with information from tooth microwear and dental calculus. This kind of analytical agreement is rare in paleodiet research. When evidence from hominin teeth is combined with skull architecture, tool use, and butchered animal remains, it is undeniable that hominin diets varied through time and across environments. Folks who hope to model their food choices after prehistoric hominins actually have myriad options to choose from!

8 Teeth as Tools, Warning Signs, and Homing Devices

Dietary behaviors recorded by tooth wear and plaque buildup are just the tip of the iceberg for dental aficionados. Alternative dental uses, technically known as paramasticatory activities, have also piqued the curiosity of scientists. In the pioneering 1963 book *Dental Anthropology*, Albert Dahlberg suggested that during prehistory, teeth "had been called on to do unusual things."[1] Yet subsequent anthropological studies—replete with images and reports of modified remains—suggest that nondietary uses of teeth were not out of the ordinary. Many of these cases highlight accidental side effects of tooth use, while in other instances dental alterations were done on purpose (the subject of the following chapter). Clinicians and parents caution against risky uses of our teeth for fear of irreparable damage, but these behaviors were so commonplace in the past that our teeth and jaws have been likened to a "third hand." All kinds of dental activities were encouraged or expected in non-industrialized societies, particularly when survival depended on it.

We'll also see below how canine teeth serve as a status symbol for certain male primates, aiding their reproductive efforts and playing into debates about the origins of human monogamy. Related studies have questioned whether modern humans have undergone "self-domestication"—leading to anatomical similarities between males and females, including males' reduced canine teeth and loss of skeletal robusticity. Teeth also provide clues about ancestral migration routes and evidence that different hominin lineages "mixed it up" genetically when their paths crossed—a subject that has captivated fiction writers, epitomized by Jean Auel and her *Clan of the Cave Bear* tales. Ancient DNA experts have recently lent some scientific gravitas to this previously implausible scenario.

The Tools Teeth Tell?

Anthropologist C. Loring Brace's perspective may surprise our aforementioned cautionary clinicians and parents, as he determined that in traditional hunters and gatherers, "the function of the front teeth … is as a clamp."[2] Cultural anthropologists have observed routine use of teeth for gripping animal skins, tendons, or plant fibers like a pair of pliers in small-scale societies. These behaviors appear to have been common in people spanning the globe from North America to Australia.[3] First-hand reports of Australian Aboriginals using their teeth to flake stone tools are astonishing to those of us who've been taught to protect our teeth above all else. Until recently, many hunters and gatherers also stripped bark and prepared digging sticks or spears with their choppers. Indigenous peoples of the Pacific Northwest may dentally tighten sled harnesses, open frozen canisters, chew animal skins to make clothing, clamp bow-drill socket holders,[4] and tear and cut seal meat. Traditional Scandinavian societies engage in similar activities, and additionally may shape tin thread and even castrate reindeer with their teeth!

Less dramatic dental habits from industrialized populations include gripping thread, nails, strings, musical instruments, and pens—a habit many of us abandoned as computers replaced our paper notebooks. When repeated over a lifetime, these behaviors may result in deep grooves, notches, chips, fractures, large scratches, and even tooth loss. One iconic example is clay pipe use, a popular way to smoke tobacco in the historic past of Europe and North America. Natural abrasives in clay often abraded the teeth used to hold the pipe, producing a circular notch that neatly matched the pipe stem. As we'll explore below, observations of this nature have been used to infer that Neanderthals also used their incisors and canines as tools.[5] C. Loring Brace was one of the initial champions of this idea, although he admitted that it isn't possible to "*really know* what prehistoric people did with their teeth."[6] We should bear in mind that using living people or nonhuman primates to understand the past means presuming that ancient behaviors have not changed much, which is a debatable assumption.

Fossils with heavily worn and beveled front teeth, such as a male Neanderthal from La Ferrassie, France, have sparked debates about the cause of his seemingly unusual dental wear. Recovered in 1909 in the Dordogne Valley, this adult skeleton includes one of the most complete Neanderthal

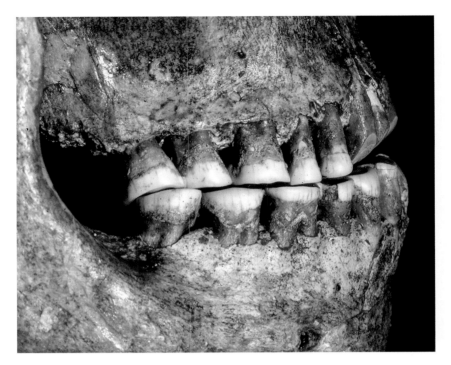

Figure 8.1
Dentition of the Shanidar 1 Neanderthal showing heavily worn beveled incisors.
Image credit: Erik Trinkaus.

skulls yet discovered. Some believe that his excessively rounded front teeth were used as a third hand or clamp, while other scientists attribute this wear pattern to having a highly abrasive diet. Subsequent discoveries, including a Neanderthal male from Iraq known as "Shanidar 1," have continued to fuel the debate. His front teeth are worn down to the roots, with inwardly curved surfaces similar to the La Ferrassie Neanderthal (figure 8.1). Shanidar 1 is one of the most elderly fossil individuals ever recovered, and he lived with severe disabilities—including the handicapping of his right arm. One interpretation is that this individual relied on his teeth in order to compensate for his unusable hand.[7] While this intriguing hypothesis is impossible to prove or falsify, as Brace noted, it is consistent with the behavior of humans living today. Examples abound of people with remarkable skill in co-opting other parts of their body, such as disabled artists who paint by holding brushes in their mouths.

Three lines of evidence have been marshaled in support of Neanderthals' use of their teeth as tools: the frequent presence of large, deep scratches; their extremely curved wear; and their large incisor and canine teeth.[8] In the first instance, Neanderthals often have large oblique scratches aligned across their anterior teeth, as would occur when an individual was holding something between the teeth to cut with a stone tool, and periodically missing the target. The scratches on Neanderthal incisors look quite similar to those on the teeth of indigenous peoples of the Pacific Northwest and Pueblo Native Americans—whom you'll recall engaged in numerous nondietary tasks. In one clever experiment, meat was clamped between porcelain tooth casts and then cut with flint tools, yielding one scratch pattern when the tool was held in the left hand and drawn to the right, and different cut marks when the tool was held in the right hand and drawn to the left. These results are similar to the pattern on many Neanderthal teeth, leading the authors to conclude that they were likely to have engaged in such behaviors and that most individuals were primarily right handed.

Another observation in support of Neanderthals' nondietary activities is the curved wear patterns on their front teeth. Objects held between the teeth that were pulled or stripped may have beveled teeth, including those of the La Ferrassie and Shanidar 1 males. Most paleoanthropologists think that Neanderthals relied on clothing to survive, particularly in more seasonal northern Eurasian environments. I had the good fortune to visit the Neanderthal Museum near the Neander Valley fossil site in Germany, featuring impressive life-like reconstructions of Neanderthals clad in heavy furs. In historic times, fur trappers treated animal skins for preservation, which included removing any muscle and fat that may adhere. Artistic renderings of Neanderthals scraping animal skins clamped between their teeth seem plausible—particularly as rounded dental wear patterns are also common in hunter-gatherers from the Pacific Northwest and Australia. Like Neanderthals, these human groups wore down their teeth rapidly and inhabited some of the most challenging environments until quite recently.[9]

A final argument in support of Neanderthals' use of their teeth as tools is the idea that their specialized facial architecture represents an adaption for this type of behavior.[10] Neanderthals have large faces with somewhat projecting front teeth that create a muzzle-shaped face, particularly without the balancing effect of a chin. I've spent a good deal of time studying enamel thickness, and it turns out that Neanderthal incisors have thicker enamel than expected from their other teeth.[11] This is similar to patterns in orangutans, who use their large incisor teeth to strip tough bark when preferred foods are scarce. Neanderthals' large front tooth crowns also have large roots, representing a possible adaption for dissipating force while working with objects held in the teeth. Some individuals' roots show thick layers of cementum, the hard glue-like substance that holds teeth in their sockets, which can thicken in response to repetitive or increased forces. These adaptations may have provided an evolutionary advantage that led to more specialized or efficient uses of this so-called third hand.

Taken together, these multiple lines of evidence represent a strong case for the "teeth-as-tools" hypothesis, although I can't help but recall C. Loring Brace's point that it is hard to be certain about ancient behaviors—particularly when conclusions rest on the use of living humans as an analogy for extinct species. Unfortunately, modern nondietary uses of teeth are often inferred from skeletal collections based on cultural accounts of similar populations. It is quite rare for scientists to be able to study the teeth of people with documented lifelong habits, which would provide much-needed tests of the relationship between particular behaviors and specific forms of dental wear or damage. While future research may overcome some of these limitations, our understanding of when and how fossil hominins co-opted their choppers to survive will likely remain somewhat speculative.[12] This is particularly true for species predating the Neanderthals, since we know far less about their culture or behavior.

Canines and Male Aggression

There's more to the story of how teeth have been used to understand hominin behavior, thanks in large measure to Charles Darwin's insightful writing in *The Descent of Man*. Darwin noticed that male and female primates differed in their tooth size, and related this to their social behavior—launching another lively debate about evolution. Next time you visit your local zoo, pay close attention to the ground-dwelling African baboons or mandrills you encounter, and watch as males yawn and casually display their impressive dental endowments. Primate canines, particularly those of certain monkeys and apes, are much taller in males than in females. Anatomical differences between the sexes of a single species are known as *sexual dimorphism*. For some primates, sexual dimorphism in canine size exceeds differences in body size. While it is natural for large primates to have large teeth, male canines are often enlarged more than other teeth, as well as the rest of the body. For example, male mandrills are approximately 3.4 times heavier than females, while their canines are 4.5 times longer, reaching an impressive height of 1.8 inches or 45 millimeters on average (figure 8.2).[13]

Canine sexual dimorphism has become a classic example of *sexual selection*, an evolutionary process that Darwin distinguished from natural selection. As he explained: "Sexual selection depends on the success of certain individuals over others of the same sex in relation to the propagation of the species; whilst

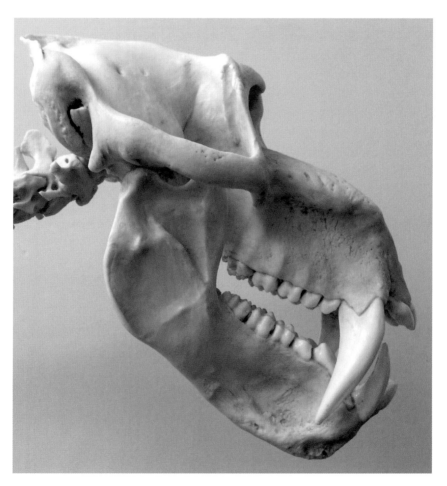

Figure 8.2
Extremely large canines in a male mandrill skull.

natural selection depends on the success of both sexes, at all ages, in relation to the general conditions of life."[14] Darwin suggested that large canines functioned as "weapons for sexual strife,"[15] meaning that large canines give male primates an advantage over smaller-fanged males when competing to mate with females. He believed that females choose mates with particular anatomical specializations, such as primates' large canines or elaborate plumage in birds, to ensure that their offspring will inherit these favorable traits and eventually reproduce themselves. These canines are not just tall points—they're dagger-like blades! Apes and monkeys have evolved a powerful mechanism for maintaining sharp edged-teeth; upper canines rub against a long, sloping lower premolar surface that serves as a biological whetstone.

Numerous scientists have tested these ideas. For example, a study of mandrills showed that males with large canines fathered the most offspring, and this happened when canines were at least 2/3 of their final unworn height.[16] Younger males with unerupted or partially erupted teeth and older males with worn or chipped teeth rarely produced offspring in this community. These results are really impressive, because it's difficult to measure the relationship between anatomical features and reproductive success in primates. Their slow growth, long lives, and promiscuous mating behavior complicate attempts to test whether specific features actually increase "evolutionary fitness." In this example, the large canines of male mandrills enhanced their owners' genetic contributions to the next generation, making them a classic evolutionary adaptation.

Darwin got it right again—at least concerning the evolution of large canines in mandrills. Does this trend explain our own lineage? Biological anthropologists have studied how canine size relates to behavior across different nonhuman primates, with the ultimate goal of understanding what early hominin social systems might have been like.[17] Mandrills and certain baboon species have polygynous social groups, in which a single, dominant male monopolizes reproductive access to multiple females through combat with male competitors. A similar trend holds for orangutans and mountain gorillas, exemplified by the iconic silverback male gorilla in charge of his harem of females and their offspring. Separated by millions of years of evolution, these despotic monkeys and apes are united by extreme sexual dimorphism in canine size and body mass. Conversely, primates with social systems in which females mate with multiple males, or those that are largely monogamous, often show less sexual dimorphism.[18]

In order to determine sexual dimorphism in ancient hominin canines, we first need to know the biological sex of individuals with measurable teeth. Forensic anthropologists have identified a number of traits in modern human skeletons that are fairly reliable for sex determination. The best predictor is the shape of the pelvis, reflecting adaptations that allow women to birth big-brained human babies safely. Other useful features include the length and thickness of the limb bones, as well as the sites where muscles attach to bones—all of which are greater in males, on average. Facial size and shape also reflect a degree of difference between the sexes, including the development of particular ridges on the skull. This exercise becomes more challenging for teeth found without other parts of the skeleton.

Modern human canines are weakly dimorphic; male canines are about 5% to 10% larger than those of females.[19] The rest of the dentition differs by less than 5%. Thus incisors, premolars, and molars aren't particularly useful for determining the sex of modern skeletons, since overlap between the sexes is so great that predictions are little more than guesswork. An important exception we'll return to later is when teeth preserve DNA, which is uncommon in fossil teeth found outside of Eurasia. In the concluding chapter of this book, I'll also describe a promising new approach that targets enamel proteins that differ between males and females. If this method can be applied to ancient hominins, we may one day be able to get a handle on how much males and females varied in the distant past, which would better inform our understanding of their social groupings. For the time being, we're forced to turn to other approaches.

Anthropologists have measured the canines of our closest living relatives in order to predict the sex of fossil individuals, reaching cautionary conclusions.[20] The main complication is that chimpanzee canines show highly overlapping dimensions between males and females, as is the case for living humans. Since our common ancestor with chimpanzees is also likely to have had overlapping sizes, attempts to infer the biological sex of canines found without other skeletal remains are problematic, particularly for canines from large females or small males. This issue came into focus with the recovery of isolated canines from *Ardipithecus ramidus*—the most apelike early hominin known at the time. The discovery of a skeleton with a lightly built skull eventually helped to clarify its canine dimorphism.[21] Dental specialist Gen Suwa and his team measured the canines from the adult skeleton, and ran a number of statistical simulations on the isolated teeth.

It turns out the skeleton's canines are smaller than most canines found so far from this species, leading the discoverers to conclude that it was female. *Ardipithecus* appears to shows a degree of canine sexual dimorphism that is intermediate between living humans and bonobo chimpanzees—a close cousin of common chimpanzees.

Paleoanthropologists now agree that early hominins and australopithecines had less canine sexual dimorphism than great apes and Old World monkeys. On the face of things, this would suggest that our male ancestors engaged in less direct competition than primates with more dimorphic canines. Owen Lovejoy, an accomplished anatomist, sparked intense debate in the 1980s by suggesting that ancient hominins adopted a monogamous or nuclear family structure, facilitating our evolutionary success.[22] Although many scholars have pointed out flaws with Lovejoy's theory, artistic renderings and museum exhibits at that time dramatized an *Australopithecus afarensis* couple strolling across a plain together. This romantic scenario was inspired by actual footprints immortalized in fresh volcanic ash at the 3.6 million-year-old site of Laetoli (figure 8.3). While the neighboring large and small fossil tracks could have been made by a male and female hominin, it is unclear whether the individuals were traveling next to each other, or even walking at the same time.[23]

Do measurements of canine sexual dimorphism support Lovejoy's idea that early hominins were living in nuclear families? Probably not. Biological anthropologist Michael Plavcan has studied primate canines, sexual dimorphism, and social systems for decades.[24] His research show that low levels of canine dimorphism do not predict primates' mating systems accurately. In other words, large differences in canine size between males and females reliably point to polygynous mating systems, but the absence of marked differences does not necessarily mean that a species is monogamous. Bonobo chimpanzees are a good example of why this is so, since male and female canine sizes are more similar to one another than are those of most other great apes, yet bonobos are highly promiscuous in their sexual behavior.[25] The key piece of information for interpreting this seeming exception is the low level of aggression within and between bonobo sexes. They are a provocative example of a polygamous primate that evolved to "make love not war"—an approach that serves to keep tensions at a minimum and reduces their need for large canines. What this means in terms of hominin reproductive behavior remains an open question.[26] Given information from living

Figure 8.3
Fossilized footprints from two different-sized hominins preserved in volcanic
ash. Reproduced from Getty Conservation Institute, *The Laetoli Conservation
Project: A Visual Summary: A Collaboration of the Tanzanian Department of
Antiquities and the Getty Conservation Institute (1993–1998)*, rev. ed. (Los
Angeles: Getty Conservation Institute, 2011). Image credit: Tom Moon.

primates and the depauperate fossil record for the initial period of human evolution, I tend to side with Plavcan's position that the pattern of sexual dimorphism in hominin canines is insufficient for determining whether our early ancestors were most like monogamous gibbons or polygamous chimpanzees.

Before we throw out the baby with the bathwater, it's worth returning to another one of Darwin's ideas. Rather than focusing on social systems, he persuasively asserted that hominin canines shrank over evolutionary time spans when humans replaced them with "stones, clubs, and other weapons."[27] Ralph Holloway reviewed this idea nearly a century later—by then, it had become a kind of anthropological orthodoxy. Holloway astutely pointed out that it wasn't clear why hominins couldn't have large canines *and* use tools for fighting.[28] Neither Darwin nor Lovejoy gave a reason for why smaller canines would be more advantageous than large ones. In contrast, Holloway proposed that canine reduction occurred as a by-product of evolutionary selection for reduced male aggression, rather than as a consequence of tool use. He hypothesized that cooperative behaviors arose in some hominin groups, giving them a competitive advantage over less cooperative groups, and that this led to a decrease in both aggression and the hormones that influence secondary sexual characteristics—including canines.

A recent model of "self-domestication" in our own species extends this reasoning by linking hypothesized changes in testosterone—the androgen hormone related to aggression and dominance behavior—with specific trends observed in the fossil record.[29] Anatomical changes that have occurred since the origin of *Homo sapiens* include decreases in ridges and crests on the outside of our skulls, as well a reduction in our brain and canine sizes. Some point to similar anatomical changes in dogs selectively bred for more tolerant behavior. I found this potential explanation for reduction in canine dimorphism to be tantalizing, so I dug into the literature to better understand how teeth change as a result of domestication and what this might tell us about our own evolution. Unfortunately, there isn't much direct evidence that canine dimorphism decreases in domesticated dogs and wolves. It's easy to get confused here, as domestication can lead to smaller teeth and dental crowding in the shortened faces of subsequent canid generations— but no one has shown that canines shrink more in males than in females.

Related versions of this theory invoke a parallel trend of reduced canine dimorphism in bonobo chimpanzees, which may have evolved from a more aggressive ancestor. Yet sexual dimorphism in bonobo canines is similar to West African common chimpanzees, which doesn't support the idea that low dimorphism is unique to the mellow bonobos.[30] Finally, theories of self-domestication in our own species do not account for the decrease in canine size since the time of *Ardipithecus* and *Australopithecus*. While it is tempting to speculate on the origins of monogamy, cooperation, or reduced aggression from measurable anatomical features, more evidence is needed to convince most paleoanthropologists. Perhaps additional research on the effects of hormones on dental development and canine size will give us more insights into social behavior over the course of our own evolution.[31]

Teeth as Markers of Place and Time

Beyond providing hints about our ancient dietary and social behaviors, fossil teeth can tell us about where and when our ancestors first spread out across the globe. In chapter 5, we established that Darwin correctly identified Africa as the place of our origins, despite the fact that hominin fossils had yet to be found there. Today we are the most widely dispersed primate on the planet, for which we can thank our cooperative cultures and complex technological innovations. Wanderlust is part of our evolutionary heritage. At least one species of *Homo* entered Europe and Asia nearly two million years ago, followed by subsequent waves of later migrants, including our own species. Scientists piece together the timing and routes of ancient migrations from fortuitous discoveries of fragmentary skeletal remains and stone tools. Important refinements in geochemistry and advances in molecular biology have also given us a new understanding of the peopling of the world over the last 100,000 years.

Before we examine these cutting-edge approaches, it's worth noting how teeth play a simple but essential role in locating our ancestors on distant landscapes at various times. For example, current evidence for the arrival of *Homo sapiens* in Asia comes from isolated Chinese teeth dated 80,000–120,000 years ago, followed by two teeth from Sumatra deposited 63,000–73,000 years ago.[32] These dates surprised some scholars, because they demonstrate that modern humans reached Asia much earlier than Europe. In addition to learning when our species spread into various regions, questions remain about what happened to those inhabitants and how they relate to the people who occupy these places today. Answers have been sought from tooth and skull shape for more than a century, and most recently, from ancient DNA.

To the trained eye, tooth crown surfaces show subtle variations, resulting in patterns that are somewhat distinct among human populations.[33] A striking example are the "shovel-shaped incisors" often found in Native Americans and some Asian groups (figure 8.4). Their upper incisors have a scooped-out look, resulting in a curved biting edge that is easy to sort from the more simplistic flat shape of incisors from Europeans and Africans. Features like shovel-shaped incisors that are hard to measure are dubbed *nonmetric traits*, meticulously documented by Christy Turner and colleagues. Turner was an academic pioneer who spent more than 40 years conducting research while training multiple generations of dental anthropologists. His careful studies of tooth shape led him to develop a theory for the colonization of the New World. Turner hypothesized that Northeastern Asian individuals migrated

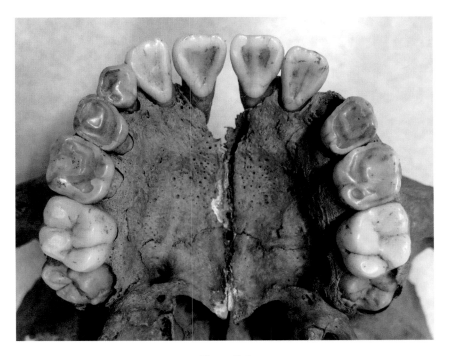

Figure 8.4
Shovel-shaped incisors (top of image) in a young Native American individual.
Human jaw courtesy of the Phoebe A. Hearst Museum of Anthropology and the
Regents of the University of California (catalogue no. 12-9499(0)).

across the Bering Strait in a series of three waves that gave rise to different indigenous peoples of North and South America. Although recent studies have questioned specific details of Turner's model, it is widely accepted that nonmetric features of tooth crowns can distinguish human populations as well as skull shapes. Anthropologists now prefer to use these characteristics in combination with information on languages and genetics to determine how human populations relate to one another. Yet sometimes dental size and shape are the only available clues for determining a prehistoric person's identity and affinity to living people.

Innovations in elemental and molecular analyses have expanded what we know about global colonization from the studies of tooth crowns that established dental anthropology as a scholarly field over 60 years ago. For example, the investigation of tooth chemistry for diet reconstruction introduced in the preceding chapter can also reveal migration histories. Isotopes, or atomic variants of a specific element, are especially powerful for exploring childhood environments. The element strontium has a characteristic isotopic signature in geological formations, varying according to age and how the rocks were formed. Strontium isotopes are naturally absorbed by plants and the animals that ingest them—including hominins. As bones and teeth mineralize during childhood, they lock in local geology like a global positioning system, or GPS. Ratios of different strontium isotopes can be measured to determine if a person or animal is native to the local environment, or if they grew up in a different area and subsequently moved to the place where their remains were found. It may be hard to imagine living in a small region for your entire life, but before the domestication of animals or the invention of vehicles, our ancestors were limited to exploring their world on foot.

Strontium isotopes have been widely used to investigate recent human remains in Europe and the Americas, and a growing number of studies have begun to sample more ancient African fossil hominins as well. I had the good fortune to collaborate with an innovative team of archeological scientists who were investigating the first Neanderthal fossil discovered in Greece.[34] By analyzing growth lines and strontium isotopes in its third molar, we found that this individual moved a few times between the ages of 6 and 11 years. Moreover, the strontium values were higher than the coastal cave site where it was discovered as an adult. The closest geological match to these values was a region approximately 20 kilometers away. This is near a prospective source of raw material for making stone tools, which may have been an important educational venue during the Neanderthal's youth.

Pinpointing the precise childhood home of this individual requires additional information on local geology, however, our study was a meaningful first step toward understanding how far Neanderthals ventured during their lifetimes. Most approaches rely on more circumstantial evidence of landscape use, such as the types of animals recovered alongside the fossils or the geological source of rocks used to craft stone tools. Ongoing studies will benefit from large-scale initiatives to map strontium signatures throughout vast regions, which could allow a sort of geographic fingerprinting of teeth. Earth scientist Malte Willmes, for example, painstakingly mapped strontium isotopes throughout France, creating an invaluable resource for inferring the childhood homes of numerous Neanderthal and modern human remains.[35] Being able to compare isotope values in prehistoric humans to a regional strontium map will also take a good deal of guesswork out of identifying immigrants, who are often otherwise indistinguishable in archeological populations. It's not hard to imagine how this could also strengthen forensic investigations of unidentified skeletal remains, although the significant time and cost have constrained modern detective work for the time being.

Several of my collaborators built on the Greek Neanderthal approach with an ambitious study of strontium isotopes in australopithecines from South Africa.[36] They learned that certain individuals appeared to be locals, while others grew up several kilometers away from where the teeth were found. Remarkably, an assortment of small teeth showed higher strontium ratios than most of the larger teeth, whose values mostly matched the local geology. This result led the authors to suggest that the small teeth belonged to females who had emigrated from their birth communities sometime after childhood. Female dispersal is common in African apes and some human societies. It ensures that related individuals do not reproduce with one another, but it may involve considerable risk to the individuals leaving their home groups. I was impressed by this study and took a closer look at information from the canines, the most reliable tooth for sex determination. Eight of these were either much bigger or smaller than the average size of a mixed-sex sample, and of these, six fit the trend described by the authors, while two did not. In the latter cases one small-toothed individual showed a local strontium signature, and one large canine appeared to be from an immigrant. Although this study was based on a small number of individuals, finding a pattern in 75% of them is certainly worth further investigation.[37]

Teeth also harbor important metronomes.[38] In the previous chapter I mentioned ^{14}C dating, which can be effective for teeth that are less than

55,000 years old. This method only works when a sufficient amount of collagen is present in dentine, and it hasn't been contaminated with modern carbon.[39] When contamination is suspected, or the collagen has been compromised by water, heat, microbial activity, or age, scientists must use other dating approaches. My colleague in the Australian Research Centre for Human Evolution, Rainer Grün, is the person paleoanthropologists turn to when they need a chronometric lifeline.

Rainer and his students have spent decades refining dating methods of last resort, which are known as uranium series and electron spin resonance (ESR).[40] He gets the call when it's impossible to date organic remains, volcanic layers near buried fossils, or burned tools at hominin sites. Uranium-series dating relies on the principle that the mineral that stiffens our skeletal tissues also passively soaks up natural uranium from burial environments. In order to estimate how long something has been buried, scientists first measure the amount of radioactive uranium and thorium isotopes that have seeped into a sample. Like ^{14}C dating, the ratio of these isotopes can provide an age for a fossil bone or tooth. Uranium series can only be used for hominin remains that date to about 500,000 years ago or less, providing a minimum age only, since bones and teeth may be buried for hundreds or thousands of years before uranium begins to turn up in measurable amounts.

Rainer and colleagues then turn to ESR dating to fine-tune these results, which is based on the fact that natural radioactivity affects teeth at an atomic level. Cosmic rays and radioactive elements in the burial environment cause the mineral component of enamel to lose electrons. These free electrons become trapped in tiny structural defects, accumulating over time in proportion to the radioactive dose that the tooth is exposed to. Determining the amount of radiation and the speed of its impact allows scientists to estimate when electrons first started becoming trapped, or how long ago the tooth was buried. This method is often used as a last resort since it involves complex modeling and assumptions that the environment hasn't been modified since the tooth was initially deposited. However, comparisons of ESR dating with other techniques often yield reasonably consistent results. Several fossils I've highlighted have been dated with ESR, including the earliest *Homo sapiens* material from Northern Africa at 300,000 years old. In the following chapter we'll learn about the oldest-known human skeleton in Australia, which was also dated with this approach.

Ancient DNA: The Ultimate Disruptor

You may have gathered by now that I'm rather partial to the mineralized nature of our teeth, largely because this is what makes them resistant to destruction while recording our development, health, location, and diet during childhood. Teeth can also serve as protective capsules for DNA for many thousands of years, especially in cool and dry environments.[41] I gained firsthand knowledge about the recovery of DNA from fossil bones and teeth while working at the Max Planck Institute for Evolutionary Anthropology, currently the global hub of ancient DNA research.

Because DNA and protein make up less than a third of the dentine when an individual is alive, finding original DNA after death can be a real challenge. Most of the genetic material recovered from human remains comes from the microbes that hasten decomposition. To combat some of these difficulties, protocols for optimizing DNA recovery are often initiated in the field even before human fossils are recovered. Imagine suiting up like an astronaut or biohazard specialist in order to climb into a remote cave to excavate ancient human remains! Protective equipment, including facemasks and gloves, minimizes the risk of contaminating fossils with the DNA of the excavators. Freshly exposed fossil samples are immediately refrigerated to minimize degradation after they are freed from the surrounding soil or stone. They are then transported to highly specialized laboratories with "clean rooms," where the bone or dentine is ground into a powder for genetic sequencing and analysis.

Paleogenomic studies are progressing at an even more rapid clip than elemental analyses, driven by breakthroughs in laboratory procedures that separate an individual's DNA from environmental contamination, as well as increases in the efficiency of molecular sequencing techniques. Ancient DNA specialists can produce an entire genome from a few hundred micrograms of well-preserved bone or tooth dentine—less than the weight of a single aspirin tablet. Forensic scientists also retrieve DNA from modern teeth to determine the identity of victims of crimes or natural disasters, although typically not with the immediate results dramatized in popular television shows like *Bones*. These cutting-edge approaches have fundamentally revised our view of the biology and behavior of ancient hominins, particularly our evolutionary cousins, the Neanderthals.

When teeth remain in bony jaws after burial, they are more protected from degradation than most other skeletal components. Moreover, the pulp space, home to many cells during our lifetime, is buffered from the external environment by largely impermeable tooth crowns as well as the root system. Recall from chapter 1 that the dentine-forming cells sit inside the pulp space after secreting the bulk of the dentine, with their long tail-like processes running almost to the border with the enamel or root surface in long, thin channels. These dentine channels also hold mitochondria, tiny energy-producing structures that contain maternally inherited DNA.

Nuclear DNA, inherited from both parents, can be only be found in the nuclei of dentine- and cementum-forming cells, making it a more rare but highly prized source of genetic information from teeth. Cementum traps its own secretory cells inside dense mineralized layers, and can yield even more DNA than dentine.[42] Finally, new research has shown that tooth calculus may encapsulate cells from our mouths, in addition to microscopic traces of our diet and the local microbial community. For example, mitochondrial DNA from six Native Americans has been recovered from calculus scrapped off their teeth approximately 700 years after they died.[43] This exciting new breakthrough will surely be attempted on key individuals housed in anthropology museums worldwide.

During my time at the Max Planck Institute, I worked with evolutionary geneticists and archeological scientists who were developing sampling protocols for ancient DNA extraction and elemental studies of Neanderthals. State-of-the-art labs fill several floors of the stunning glass-walled institute, which has welcomed curators in charge of some of the most important fossils in Europe. Scientists whose work relies on taking samples for study know that custodians often loathe being asked to allow physical alteration of a fossil, since they are responsible for ensuring that scientific study is balanced by the preservation of these invaluable relics.

Studies of a Belgian Neanderthal from the site of Scladina illustrate the potential payoff.[44] Though the research did require removing some tissue from two teeth, we have recovered enamel proteins, which are like those of modern humans, as well as some of the oldest mitochondrial DNA sequences from a Neanderthal at that time. This individual had an unexpectedly rapid dental development and a premature weaning event at 1.2 years of age (figure 7.3) before it died at 8 years old. These insights into Neanderthal development, evolution, and behavior could not be gleaned from any other method of study.

Mitochondrial DNA and sex chromosomes have been retrieved from a group of 12 Spanish Neanderthals who are thought to have died together and been cannibalized by other Neanderthals.[45] Three adult males in the group descended from the same mother, while three adult females had different mothers, suggesting that males remained in their natal group while Neanderthal females dispersed into new groups prior to adulthood. This is similar to the aforementioned study of strontium isotopes in South African australopithecines that pointed to a local "band of brothers." The Spanish Neanderthal group had low genetic diversity, implying that they were fairly isolated from other prospective mates.

One of the most impressive advances in paleogenomics is the recovery of sufficient DNA to reconstruct the complete genomes of hominins from the Denisovan Cave in Siberia.[46] Two large molar teeth, one baby tooth, and one tiny finger bone have been sampled from a relative of the Neanderthals that has yet to be given a formal species name. Incredibly, they are the only identified remains of an ancient group that coexisted with both Neanderthals and our own species. We know more about the Denisovans' genomes than we know about how tall they stood, how big their brains were, or whether their anatomical features were different enough to recognize them as a new species. Amazingly, some of the Denisovans' DNA lives on in modern humans, particularly people from Eastern Asia, New Guinea and nearby islands referred to as Melanesia, as well as Aboriginal Australians. As the ancestors of these modern groups left Africa and moved through Eurasia and Southeast Asia, some of them interbred with the Neanderthals and Denisovans, carrying their genes all the way to Australia!

I hope I've convinced you, and a few curators-in-training, that these fascinating insights justify the removal of bits of bone and teeth in select cases—especially when scientists carefully photograph, mold, and CT scan the fossils first. As more of our analytical tools and knowledge moves into "the cloud," the production of genomic libraries and high-resolution virtual 3D models preserve fossils in ways that make them even more accessible to the scientific community than costly visits to simply gaze upon the original remains.[47]

It certainly can be challenging to go from speculating on ancient behaviors to identifying concrete evidence in the fossil record. Natural experiments or observations of people in certain occupational conditions may strengthen our understanding of how teeth may have once been versatile prehistoric tools. For obvious reasons, it is unethical to experiment on human subjects in ways that would permanently alter their teeth. Scholars are often left with little recourse but to infer behavior from what is known about living human populations or nonhuman primates. A similar logic has guided interpretations of ancient dietary evidence from studies of tooth microwear. We'll see in the following chapter how one of the most widespread prehistorical markers of hygienic behavior has been confirmed through creative experimentation.

Other inferences that inspire debate and elicit broad interest, such as the origins of human monogamy, require additional validation. Unique features of the human animal, including our large brains and complex cultures, complicate gross generalizations from nonhuman primates. Elemental and genetic information locked inside tooth crowns and roots may yield the most specific behavioral records of our ancestors, such as how their social groups were organized and whom they mated with. Interpretations of DNA from ancient hominins and prehistoric humans over the past decade have electrified the field of paleoanthropology.[48] Scientists now acknowledge that our biology has been significantly influenced by intimate encounters with other hominins. Some have gone so far as to suggest that genetic contributions from these lineages facilitated the evolutionary success of *Homo sapiens*.[49] Whether or not we are comfortable considering ourselves "hominin hybrids," evidence from ancient teeth has illuminated the dynamic behavioral context in which our species evolved.

9   Tooth Manipulation through the Ages

Why do humans seek out elaborate ornamentation like gold caps, grills, and jewel implants for our choppers? What motivated prehistoric humans throughout the world to file and notch their teeth? The odds are that it started with a little discomfort; after all, pain is a great motivator. Early dentists first co-opted simple stone tools to excavate cavities, while more ancient hominins crafted toothpicks from plants. An even more dramatic example is the prehistoric practice of trephination—intentional removal of cranial bone. Modern anthropological museums house skulls that demonstrate the impressive techniques of Stone Age surgeons. This risky procedure is thought to relieve the internal cranial pressure that can build up after physical trauma.[1] Numerous patients survived being trephined, growing new bone to fill holes that left their brains exposed under thin meningeal layers. Others were less fortunate. Skulls bearing sharp-edged cut marks without the reparative bone our body quickly forms reveal that cranial surgery was often fatal. In the middle ages, trephination was also employed as a supernatural or psychological treatment to free individuals from demonic influences or aberrant behavior. This technique has even captivated a handful of contemporary people, who've gained notoriety by cutting holes in their own skulls for the purported spiritual or psychological benefit. As we'll discuss in this chapter, similar motivations may have led to widespread tooth modification and even complete removal, although—for better or worse—the recent spread of Western standards of beauty has curtailed these intriguing ancient behaviors.

Stone Age Dentistry

Cranial trephination may well be the oldest form of skeletal surgery, but dental interventions followed shortly after—beginning at least 14,000 years ago.[2] Compelling evidence comes from an adult male skeleton found in an Italian rock shelter. This wasn't a routine prehistoric burial, as he had been interred with a bone and stone toolkit, then covered with stones painted with red ochre. The use of this natural pigment and inclusion of weaponry indicates that this was someone special, and evidence from his teeth has recently cemented his place in history. During a careful examination of his skeleton, researchers noticed that a pit on the chewing surface of his third molar had chipped enamel and deep scratches inside. Several cavities had formed inside and around the pit, leading to a significant loss of enamel and painful exposure of dentine. The scratches resembled cut marks commonly found on butchered bones, implying that a sharp tool had been used to excavate the pit.

In order to determine which type of tool produced the marks, the team set about "treating" modern human teeth with small wood, bone, and stone tools from this time period. The stone tools produced striations very similar to those in the Italian male's tooth, indicating that they were likely used to attend to the diseased region. Importantly, the scratched and chipped enamel near the outside of the pit was polished after the surgery, while the deeper areas retained sharp-edged scratches. It appears that the patient survived the treatment and continued chewing for some time afterward. The team hypothesized that this surgical intervention represented an adaptation of toothpick use, a behavior I'll discuss in the following pages that was quite commonplace by this point in our evolutionary history.

As we learned in chapter 3, the development and spread of agriculture and industrial sugar production have driven epidemics of cavities in modern people. Ancient dentists developed increasingly sophisticated surgical practices as agriculture, and cavities, became more common. For example, teeth from a graveyard in Pakistan buried 7,500–9,000 years ago show clear signs of dental work.[3] Several molars have conical holes in their chewing surfaces that are a few millimeters wide and deep—similar to holes produced by the tip of a pen or a small screwdriver. Careful microscopic investigation of their insides revealed circular rings produced by a rotating tool, as occurs during drilling. The researchers also experimented to understand how these holes were produced, testing the efficacy of flint drill heads found in archeological sites. Pointed stone bow-drills were often used for bead production—and it turns out that this precursor to modern dental drills is quite effective! In less than one minute of manual rotation, flint-tipped drills can produce circular holes in modern teeth that resemble those found in the prehistoric individuals. These interventions allowed the Pakistani patients to use their teeth afterward, as the margins of the drilled holes showed more smoothing that the deeper parts, similar to the Italian male's molar. Our pioneering dental surgeons smartly repurposed commonplace tools and techniques to alleviate oral discomfort. As you'll see below, they also created remarkable fashion statements and elaborate symbolic ornamentation.

Modern dentists don't simply drill holes in teeth and send people on their way; they've developed numerous approaches to protect fragile crowns and highly sensitive roots. Without access to refined metals, ceramics, or plastic resins, Stone Age dentists turned to naturally occurring materials as salves. A 6,500-year-old lower jaw of a male unearthed in Slovenia shows such evidence of a palliative treatment.[4] His worn and fractured canine tooth appears to have been covered with beeswax, which may have sealed the fracture and alleviated some of the discomfort. In this case, beeswax does not record microwear like tooth enamel, rendering it difficult to be certain that the application occurred before death, and if so, how long. The 14,000-year-old skeleton from Italy was also buried with a wax-like substance near his skeleton, and the pit in his molar contained an organic residue, although scientists could not determine what it was made of. Organic products are notoriously difficult to identify in prehistoric contexts, since microbial attack leads to their decomposition and loss in most cases. It's quite likely that humans used natural products to try to relieve pain during antiquity, but we need to recover more of these brave dental patients to clarify whether they actually received any relief.

When dental decay or infection reaches a critical point, modern clinicians remove diseased teeth, leaving holes in the gum and bone that will eventually heal and close—provided the infection clears up. Tooth removal is a point of no return, as our teeth will not naturally replace themselves, and the remaining dentition often works differently after teeth are lost. Were our prehistoric dentists extracting diseased teeth as well as scraping and drilling tooth crowns? Unfortunately this question is nearly impossible to answer before recorded time. During advanced stages of dental disease, the tissue attachment system weakens, and loose teeth can detach with only minor pressure. Natural loss may also be hastened by digital manipulation, a technique many children and even chimpanzees have learned as they wiggle their baby teeth. The bottom line is that it is difficult to determine when a diseased tooth has been intentionally removed rather than being lost naturally.[5] Teeth I've studied from modern dental clinics often retain signs of their extraction; microscopic scars of metal forceps or sharp probes hint at their painful release. Yet without similar evidence of ancient interventions, scientists are unable to pinpoint the origins of extractive procedures. In the final chapter of this book, we'll consider how breakthroughs in stem cell research may one day make the seeming irreversible loss of teeth a thing of the past.

A close cousin to prophylactic extraction is tooth *avulsion*—the intentional removal of healthy teeth, which makes a dramatic appearance several thousand years ago.[6] Traditions in modern small-scale societies can be helpful for recognizing and interpreting behaviors in prehistoric people, particularly for activities that are foreign to those of us who decipher the bony evidence. Cultural anthropologists report that humans avulse teeth across the globe. In many cultures some or all of the incisor teeth are removed, and baby teeth may also be avulsed prior to their natural loss. Tooth removal may take place at different ages, and at multiple times through the life course. I'll admit that I find accounts of avulsion difficult to read. Individuals are typically not provided with any anesthetic, and in some cases the patient is expected to remain unexpressive while an iron spike, thin knife, or an implement made of a stick and rock is used to dislodge their tooth crowns and roots. Those of us who wince at the mere thought of getting our teeth cleaned have truly met our match. We spend considerable sums to protect and retain our healthy teeth, and can't imagine voluntarily giving them up!

Why would people remove teeth that are disease-free? Some scholars view it is a form of body modification akin to tattooing, piercing, or head shaving. Many of us adorn ourselves in various ways to express our cultural identity or group affiliation, as well as our status and gender identity. The reasons for tooth avulsion vary across cultures, and include aesthetic value, symbolic intimidation, and initiation and mourning rituals. It appears to confer a social benefit or hold personal meaning that is significant enough to justify the pain and risk of infection, as well as the loss of function. Other reasons put forth include language production, tool use, basket making— and even to allow unfortunate individuals with lockjaw to eat.

Some of the earliest definitive evidence for avulsion comes from skeletal remains of African hunter-gatherers dated 13,000–15,000 years ago, which include numerous adults with high percentages of avulsed front teeth.[7] This tradition may have originated in the Northwestern African region known as the Maghreb, spreading throughout Africa over a few thousand years—where certain tribes continue avulsing teeth to the present day. However, skeletal material from ancient Japan, Southeast Asia, and Australia points to the intriguing idea that tooth avulsion began independently at different times on several continents.

We've already seen that it can be difficult to distinguish intentional tooth removal from loss due to disease. Complicating things further is the fact that incisors and canines have a simple tapered root, making them prone to fall out of skeletonized jaws after death. Thus a novice observer could easily mistake postmortem loss or congenital absence for intentional avulsion. This has led dental anthropologists to develop specific criteria to identify avulsion in skeletal remains. Typically, a deceased individual is thought to have undergone avulsion if the bone around the missing front tooth (or teeth) does not look diseased, if there is symmetry in the absence of corresponding right and left teeth, and if the alveolar bone surrounding the missing teeth had begun to regrow. Instances when this pattern is present in multiple individuals in a population are even more convincing.

These criteria can help us to evaluate a case of possible avulsion in prehistoric humans from the Willandra Lakes region in Australia.[8] One individual who lived approximately 40,000 years ago appears to have lost both lower canines simultaneously and lived for a considerable period afterward. This individual, known as Mungo Man, is currently the oldest human skeleton recovered from Australia.[9] This Aboriginal individual was found covered with red ochre brought from a distant source—suggesting that he was a significant social figure. A second individual from the Willandra Lakes is missing both lower central incisors, and this person also lived long enough for the bone to have filled in completely. Their original scholarly description by Steve Webb leaves open the possibility that the missing teeth may have been intentionally avulsed. Although these two individuals possess several key indicators of tooth avulsion, the antiquity of this remarkable behavior would be reinforced by the discovery of additional remains from this region with similar patterns of missing teeth.

Ready-to-Wear Dental Ornaments

You're probably more familiar with forms of modification that are experiencing a renaissance in contemporary culture—including the popularization of "tooth jewelry." Public figures Mike Tyson, Madonna, Justin Bieber, Beyoncé, and others have famously donned gold caps and grills as part of a trend that some people believe began in the 1980s with American rap stars. Similarly, young Japanese women have embraced a cosmetic procedure that misaligns and exaggerates their canines, which they believe makes them more attractive to men. Yet anthropologists could point out that similar forms of tooth modification have been in vogue across the globe for thousands of years.[10] While pain may have motivated the earliest patients to seek rudimentary dental care, there is little question that vanity and conformity has influenced subsequent manipulations.

The most widespread forms of "dental art" practiced by prehistoric and recent people are filing and notching, resulting in a diverse array of patterns (figure 9.1).[11] In order to create such an arresting look, parts of the front teeth are removed and shaped with metal tools, hard minerals, or abrasive powders. A quick query of modern tooth filing on the Internet yields anecdotal reports, amateur images, and even videos shot by tourists visiting the Indonesian island Bali. The Balinese believe that blunting the front teeth diminishes negative emotions like anger, lust, and greed, while enhancing attractiveness. This practice rarely leads to serious health complications, particularly when the underlying pulp remains protected. For adolescents in Bali, tooth filing is an important rite of passage into adulthood.

Tooth filing and notching have also captivated historians of the New World, particularly those in search of clues about the migration of people and transmission of culture before Europeans arrived. Anthropologist Javier Romero carefully studied more than a thousand modified teeth at the National Museum of Anthropology in Mexico City, establishing a detailed classification system in 1970 that is still in use. Romero believed that filing and notching developed independently in the Americas, beginning approximately 3,400 years ago in what is now modern-day Mexico and spreading north and south over the following centuries. Artifacts such as ceramic effigies depict deities and important individuals with filed or notched teeth, pointing to a potential spiritual or religious significance for tooth modification. These elaborate designs were part of a rich cultural period of

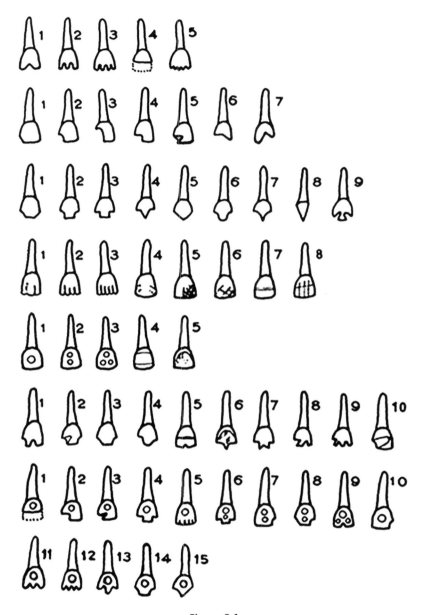

Figure 9.1

Forms of tooth modification in human remains from Central America. From
Javier Romero, "Types of Dental Mutilation in the Americas," in *Handbook
of Middle American Indians*, vol. 9, eds. Robert Wauchope and T. Dale Stewart
(Austin: University of Texas Press, 1970), 50–67. Reproduced by permission
of the University of Texas Press.

symbolism, sculpture, and architecture that attracts thousands of tourists to remote Central and South American jungles annually. The somewhat less remote Peabody Museum of Archaeology and Ethnology at Harvard University—my scholarly home for several years—has some impressive examples of modified teeth in their Mayan collections (figure 9.2).

African people have also been filing their teeth for thousands of years, which continues to this day in a number of tribes across the continent.[12] Local traditions have led to distinctive geographic variants, which may differ in the number of teeth involved, the style of filing, and the sex or age of individuals so groomed. Some tribes accentuate the pointedness of their teeth as a symbolic invocation of the power or courage of African carnivores and their prey. Like the Balinese, numerous groups modify their teeth as a marker of tribal affiliation or a means to attract sexual partners, filing routinely to mark the end of puberty. In certain communities it is more common for women to have their teeth filed, commencing from the beginning of their childbearing years.

Evidence for more ancient African practices comes from four individuals with filed teeth who are believed to be female. Radiocarbon dates from charcoal and associated plant remains indicate that these women lived 4,200–4,500 years ago. Current inhabitants of this region of Mali do not file their teeth, a change that likely reflects the complex movements of modern people or the natural turnover of cultural practices over time. More recent skeletal remains with characteristic African filing styles have been recovered in the Americas. These have been used to reconstruct details of the African slave trade. For example, five cases of tooth filing were found in a large burial population in Barbados, and subsequent strontium isotope analyses of their enamel suggest that these individuals were brought to the Caribbean from Africa before they reached adulthood.[13] This combination of characteristic filing styles and elemental signatures is a powerful example of how dental detective work can illuminate a difficult aspect of recent history.

Other impressive modifications include tooth inlay and coloring. *Inlay*—the permanent incrustation of objects on the outer surface of front teeth—appeared in the archeological record of Central America several thousand years ago (figure 9.3).[14] It seems to have developed shortly after filing and notching, and like these other forms of ancient dental art, we think that inlay yielded aesthetic value and signaled group affiliation. The use of jewels is most common in Central America, while the pliant metals gold and brass

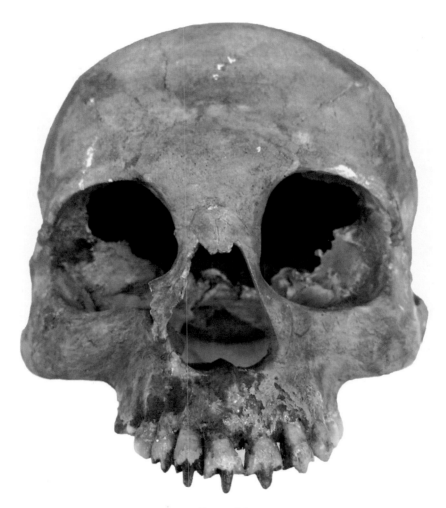

Figure 9.2
Human adolescent with filed incisors and canines from a Late Classic Period Mayan
site in the Yucatan (Mexico). The back of this individual's skull (not shown) has
deformation similar to that produced by habitually strapping a young child to a
cradleboard. Human skull (94–49–20/C2217.0) courtesy of the Peabody Museum.
Copyright 2018, President and Fellows of Harvard College.

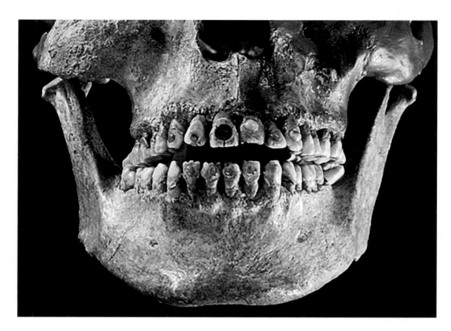

Figure 9.3
Tooth inlay in a human dentition from Central America. Image from the Mexican
National Institute of Anthropology and History.

were preferred in Southeast Asia, giving rise to some fantastically bedecked
dentitions. Simple hand drills were used to first indent the tooth, and gem-
stones like jade or turquoise were then set and bonded with an organic
resin to hold the inlay in place. In certain cultures, inlays may have been
reserved for those with high social status. Archeological excavations in the
Philippines have turned up small numbers of individuals with gold pegged
or plated teeth, in contrast to the more common recovery of people with
filed or stained teeth.[15] Additional evidence for their function as status sym-
bols includes the common presence of elaborate artifacts buried with people
whose teeth were inlayed. Like our 14,000-year-old Italian dental patient,
these dentally endowed folks were often interred with art, tools, and even
animal remains. Thus the presence of tooth inlays gives us a rare clue to
social stratification, or how humans regarded one another in the distant past.

Tooth modification also has a rich history in Asia. This is perhaps most
exemplified by intentional coloring or staining, a widespread tradition until
the past century.[16] Thomas Zumbroich has extensively pieced together its

history and the specific plants and minerals used to stain teeth. He details how organic sources were sometimes mixed with metals to produce a paste or resin then applied by hand—producing red, brown, or black teeth. Some of these coloring agents required daily reapplication, while others created a dramatic permanent effect on the enamel. As Zumbroich poetically states, "human teeth are an alternative canvas to skin, upon which differences can be inscribed as a way of defining individual and cultural identity."[17]

Tooth blackening was likely to have once been favored by many cultures, particularly in the islands of Southeast Asia. This can be detected over 4,000 years ago in the Philippines, and possibly earlier on the island of Flores. As is the case with tooth removal, it can be difficult to distinguish between intentional modification during life and environmental impacts after death. Chemicals in burial environments, or even objects like coins placed in the mouth of the deceased, can mimic the discoloration of teeth during an individual's lifetime. Moreover, the popular modern habit of chewing Betel nuts—a widely used addictive stimulant in Asia—inadvertently leaves reddish-brown stains on teeth. Scientists are careful to consider these possible complications when interpreting behavioral clues from dental remains.

What might inspire people to go to the trouble of routinely darkening their teeth? During recorded history this served as an act of beautification that helped individuals secure a mate, mark the maturation of children, and signal their group identity. Part of the appeal in some cultures also related to the diminished appearance of canines, as their pointed shape invoked a negative association with animalistic fangs, leading to a desire to hide them. Other accounts from Southeast Asia suggest that tooth coloring may have been practiced to prevent disease, including fending off the mythical "tooth worm." Ironically, a few of the plants frequently used to blacken teeth appear to have antimicrobial activities, and common binding ingredients in coloring agents such as coconut oil and citrus leaves were also used to treat toothaches.[18] Rigorous assessments of the health benefits of tooth coloring are very limited, as is the case for most accounts of traditional plant use for natural remedies. This brings us to the present, when Western ideals of beauty have led to the almost complete cessation of tooth staining. Unless you're dressing up as a ghoul for a Halloween party—an ideal time to blacken a few teeth—the primary form of tooth coloring now favored worldwide serves to lighten rather than darken teeth.

Dental Decorations

While humans have been intentionally drilling, removing, and adorning our pearly whites for thousands of years, we have been collecting the teeth of deceased humans and other mammals for an even longer period. And apparently, we did so to make a fashion statement. Members of our species have been carefully stockpiling and piercing natural objects like marine snail shells for nearly 100,000 years. Personal ornament production became widespread in the archeological record by about 40,000 years ago, when modified animal teeth appear with some regularity along with a diverse array of sophisticated bone and stone tools. These select teeth are identified from holes drilled in the single root of canines or incisors—probably so that they could be hung like a pendent, or perhaps strung together like beads. In other instances, roots were scored to create deep circular grooves that allowed them to be tied with fibrous string for suspension. Any potential ties or tethers have long since decayed, leaving some uncertainty about the exact use of these items. While many scholars assume that pierced teeth were directly worn as jewelry, members of contemporary small-scale societies often hang teeth and other ornaments from baskets, bags, blankets, and even residential structures.[19]

Preferred decorative materials share aesthetic properties; ivory, amber, tooth enamel, and mother of pearl, for example, are all lustrous in visual appearance and smooth to the touch. Opinions vary about whether there are any clear stylistic or cultural trends through time. Pierced teeth started turning up at fossil sites in Eurasia and Papua New Guinea from 28,000–45,000 years ago.[20] Randall White, an expert on prehistoric personal ornaments, has argued that certain animals were specially chosen for symbolic use. He demonstrated that the first pierced teeth were not sourced opportunistically from prey items, including the remains of animal species found with signs of hunting or scavenging. Common ornamental teeth include those of foxes, wolves, red deer, and bears, but their discarded bones are infrequently found in ancient human dwellings. Teeth may have been collected from skeletal remains found in local habitats. I'll admit to being fascinated by animal skeletons I discovered while roaming forests in upstate New York during my childhood—bringing home the odd raccoon skull or deer jaw to add to my tiny anatomical museum. Perhaps there is an innate tendency in us to collect and display striking natural items like shells, bones, and teeth.

Any guess as to what the most popular dental ornament was? The award goes to the canines of red deer—a common adornment of early European humans.[21] Pierced deer canines seem to have been so important that they were even copied and sculpted from ivory and stone. We really have no idea what made these teeth so special. Humans also branched out in their biological source material beyond four-footed mammals.[22] They collected and used the throat teeth of fish in Europe, as well shark teeth in Papua New Guinea. Prehistoric people even co-opted each other's teeth for personal ornamentation! Several archeological sites contain human teeth that show modifications similar to those of pierced animal teeth. Some are marked by deep cuts, implying that they may have been removed before their owners had become fully skeletonized. In one instance, an enigmatic lower jaw from a French cave has elicited similar grisly interpretations. Recovered with numerous modified animal teeth and an adult human's pierced incisor, this young hominin's jaw has scratches near the site where the tongue muscles would have attached to the bone. We can't be sure if the marks on the jawbone resulted from cannibalism, preparation for burial, tooth extraction for symbolic practices, or environmental modification after death. Evidence from additional fossils might help us understand why and how often human teeth were used as symbolic objects.

One of the most contentious issues in the study of personal ornaments is the question of who made them. Pierced teeth first appear in Eurasia during the period that coincides with the end of the Neanderthal reign and the initial appearance of modern humans. These groups overlapped by a few thousand years, although the pattern of replacement or potential synchrony varies from site to site throughout Europe and the Middle East.[23] One approach to figure out which hominin was present at each site is to look at cultural artifacts, which are more common than bones or teeth. Classic tool types associated with more ancient Neanderthal bones, as well as a brief and unusual artifact style that most anthropologists believe was also made by Neanderthals, disappear 39,000–41,000 years ago across a number of different sites in Europe. This has been interpreted as a sign of their extinction. Novel technological developments originally thought to be the work of modern humans first appeared around 44,000–47,000 years ago, including sophisticated stone blades, various personal ornaments, and symbolic art.[24]

The French site known as Arcy-sur-Cure is particularly germane to the question of who made the first personal ornaments.[25] Archeologists amassed an extensive collection of perforated teeth, personal colorants, and bone tools over years of excavations. These artifacts appear in the same geological layers as teeth that show Neanderthal features. We discussed in chapter 5 how dental remains—when available—are a powerful tool for identifying particular hominin species, although this too is not without difficulty. Neanderthal and modern human teeth look pretty similar, which is especially true when their teeth wear down, and anthropologists are often forced to judge subtle shape differences or internal features like enamel thickness when deciding who the teeth belong to.

The Arcy-sur-Cure teeth have special crests on the chewing surfaces, in addition to long tooth roots, which align them more closely with Neanderthals than with modern humans. Critics of the idea that the Neanderthals made symbolic items point to problems with the dating at this site, as well as the possibility that modern human cultural artifacts may have been mixed into the older layers from those above. More recent studies, however, have demonstrated that several of these artifacts were originally deposited alongside the Neanderthal teeth and are 41,000–45,000 years old—which precedes the extinction of Neanderthals.

Some scientists have speculated that similar cultural traditions are the result of encounters between Neanderthals and modern humans.[26] Paleoanthropologist Jean-Jacques Hublin and his colleagues believe that modern humans introduced their symbolic behaviors to Neanderthals in Western Europe, as represented by the personal ornaments and Neanderthal remains at Arcy-sur-Cure. Yet new evidence reveals that Neanderthals were making sophisticated bone tools 50,000 years ago in France, and piercing marine shells 115,000 years ago in Spain—long before the documented arrival of modern humans.[27] This raises three interesting possibilities: that Neanderthals and modern humans developed similar cultural traditions independently, that modern humans arrived in Eurasia earlier than the current evidence supports and subsequently influenced Neanderthal technology, or that Neanderthals introduced their culture to modern humans after the latter group arrived in Europe.

Indisputable evidence for meetings between Neanderthals and modern humans in Europe is hard to come by. Initial genetic exchange has been estimated to have occurred around 50,000–75,000 years ago, as the lineage of modern human explorers who ventured out of Africa through the Middle East incorporated some Neanderthal DNA into their genomes. Direct contact with Neanderthals occurred again when later humans moved into Eastern Europe. We know this because DNA from a 37,000- to 42,000-year-old modern human from Romania reveals the presence of a Neanderthal ancestor in his family four to six generations before he was born.[28]

Until recently, Neanderthals were thought to lack the capacity to engage in abstract thinking and extensive symbolic behaviors. Similar objections were once raised about the possibility that they made and used toothpicks— a routine behavior whose origins are now known to predate Neanderthals by more than a million years. The emerging European archeological record during this key transition between Neanderthals and modern humans currently points to more cognitive similarities than differences.[29] Returning to the example of perforated animal teeth, these objects have yet to be found with early modern humans in Africa.[30] The earliest pierced teeth and sophisticated bone tools in France predate the arrival of modern humans by several thousand years, which implicates Neanderthal craftsmen. Some scholars are reluctant to embrace yet another shift in our understanding of what does—or doesn't—make the human species unique. The long-standing idea of human superiority may reflect a subjective bias of European scholars, as well as a lack of evidence for symbolic behaviors by other hominins until quite recently.

What might have driven Neanderthals to collect and modify teeth? Some anthropologists believe that late-surviving Neanderthals made personal ornaments to maintain a sense of group identity in the face of increasing competition from other Neanderthal groups or modern humans.[31] One difficulty in assessing this idea is the lack of clear burials during this time. The fossil record during the overlapping occupation of Europe is sparse—limiting understanding of how dental ornaments were used, or what this says about the individuals who made or possessed them. Artifacts and skeletal remains from late Neanderthal and initial modern human sites are typically recovered without the kind of information gained from more recent humans, such as the 14,000 year-old Italian male who had undergone dental surgery. Recall that he was buried with a toolkit and interred under painted rocks, providing a more complete portrait than would have been possible had only a few of his teeth or tools been recovered.

Discoveries from the last few thousand years of prehistory yield more satisfying insights into the significance of pierced teeth. One of my favorite examples is an adult male Australian Aboriginal skeleton found with an impressive collection of pierced Tasmanian devil teeth.[32] His skeleton remained undisturbed after he was buried several thousand years ago, and careful excavation revealed a line of teeth encircling the neck vertebrae. This single-stranded necklace was made of more than 159 pierced teeth from a minimum of 46 Tasmanian devils. Additional tooth fragments recovered from the burial could not be positively identified, but their presence implies that the original necklace could have contained teeth from more than a hundred of these intimidating carnivorous mammals. Considerable efforts must have been devoted to collecting the teeth, piercing them, and crafting the necklace, highlighting the man's importance.

Another impressive example of ornamentation is the sensational "dog-tooth purse" from Europe, dubbed the world's oldest purse.[33] German archeologists recovered what appears to have been a bag studded with several hundred tooth crowns. After more than 4,000 years, the bag has mostly disintegrated—leaving behind neat rows of mineralized canid ornaments. While this purse is unlikely to set a trend in the contemporary fashion world, decorative natural "bling" ranging from shark tooth pendants to raptor talon necklaces are readily available from online jewelers and beachside tourist shops, pointing to their timeless appeal.

Ancient Oral Hygiene

In the previous section, I mentioned an even more ancient behavior that leaves a permanent record on teeth. Toothpick use may have begun nearly 2 million years ago, making it the earliest known dental manipulation. Scientists have identified two African hominins separated by more than a thousand miles with clear horizontal grooves on their teeth (figure 9.4). They belong to *Homo habilis* and *Homo erectus*—some of the earliest members of the genus *Homo*.[34] Are you ready to believe that long, thin grooves on the sides of fossil teeth are caused by toothpick use? It's an idea that has been debated since a European dentist first proposed it in 1911. A few decades later, the famous paleontologist Franz Weidenreich rejected the suggestion that ancient hominins used toothpicks, arguing that it "seems too grotesque to be true."[35]

Thankfully, biological anthropologist Leslea Hlusko has recently helped to settle the issue.[36] She designed a clever experiment using natural grass-stalk toothpicks, along with skulls from a baboon and a human. As we discussed in chapter 7, grass contains tiny hard particles that create microscopic scratches, and over a lifetime of chewing they wear away tooth

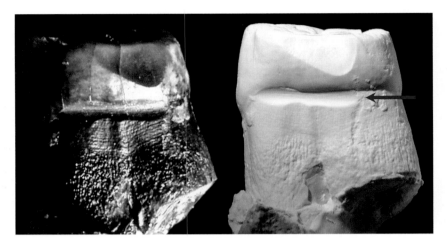

Figure 9.4

Original fossil (left) and replica (right) of an early *Homo* tooth showing a horizontal notch at the crown base believed to be caused by toothpick use. Originally
published in Leslea J. Hlusko, "The Oldest Hominid Habit? Experimental Evidence
for Toothpicking with Grass Stalks," *Current Anthropology* 44, no. 5 (December
2003): 738–741. Reproduced with permission from Tim White.

enamel. Before beginning her experiment, Leslea fitted the bone around
the teeth with artificial gums, soaked the jaws for a day, and then kept them
moist during subsequent toothpick use. Starting with the baboon skull, she
"toothpicked" with short strokes for 8 hours, which made a long, thin horizontal depression on the surface. After proving that simple grass sticks were
effective at creating grooves on a baboon tooth, Leslea turned to the human
jaw and set to work. This yielded a long, thin, horizontal groove almost
identical to the 2-million-year-old *Homo habilis* fossil. She explained to me
that she toothpicked the jaws while at an academic faculty retreat—a rather
dry event perfectly suited for doing some science! Similar grooves on the
teeth of *Homo erectus* from Ethiopia, the Republic of Georgia, and China, as
well as *Homo heidelbergensis* and *Homo neanderthalensis*, show that toothpick
use was widespread long before our own species appeared on the scene. In
fact, toothpick use may be the first hygienic habit we can detect in the fossil record.

However, not to be outdone by their human cousins, great apes have
also been seen attending to their teeth with small sticks. Bill McGrew, Caroline Tutin, and Jane Goodall documented an astonishing case of dentistry

in a group of captive chimpanzees at Tulane University.[37] This community had been relocated to Louisiana from their birthplace in Africa, and had lived comfortably together for a few years before things started to get interesting. First, a young female named Belle was observed manipulating her own loose baby teeth, which other youngsters in the community starting doing as well. When she began to employ sticks to do so, several individuals followed suit. One day Belle was seen using a cigarette-sized pine twig to groom the teeth of a young male named Bandit. Astonishingly, Bandit voluntarily submitted to Belle's care, and others gathered around to peer into Bandit's mouth and watch Belle in action (figure 9.5). Eventually Belle managed to dislodge one of Bandit's baby teeth, which she collected from him after a few minutes. The keepers eventually succeeded in distracting them long enough to recover the dental implement and the tooth—one-of-a-kind mementos for the tooth fairy!

Chimpanzees have long been known to use tools, carefully selecting and preparing thin sticks for "termite fishing," even caring for "stick dolls."[38] So it's not too surprising that they might use sticks to manipulate their own teeth. Perhaps more remarkable is the voluntary submission of a chimpanzee for intimate tending by another individual, an anxiety-producing experience for many human dental patients! Yet chimpanzees and other primates engage in extensive social grooming, spending hours carefully combing one another's fur with their hands, teeth, and lips. Dental grooming, with or without sticks, often occurred during longer full-body bouts in the chimpanzee colony. This routine social bonding likely paved the way for Bandit to relax and submit to Belle's curious dental ministrations.

If you've ever had the hull of a popcorn kernel stuck between your teeth or under your gums, you know all too well that acute and distracting sensation of oral irritation, a form of sensory awareness we appear to share with apes and monkeys. In order to alleviate this discomfort, we pick or floss our teeth—a behavior also employed by our primate relatives.[39] For example, a captive female macaque monkey was known to routinely floss her teeth with her own body hair, as well as the hair of individuals she was grooming. Her partners didn't always take kindly to her fastidious oral hygiene, periodically objecting when she pulled too hard on their fur! Similarly, free-living macaques in Thailand will remove human hairs while riding on the heads of tourists, which they may then use to floss their teeth. Mothers even

Figure 9.5
Adolescent female chimpanzee Belle grooming Bandit's teeth. Originally published
in William C. McGrew and Caroline E. G. Tutin, "Chimpanzee Dentistry,"
The Journal of the American Dental Association 85, no. 6 (December 1972):
1198–1204; William C. McGrew and Caroline E. G. Tutin, "Chimpanzee Tool Use
in Dental Grooming," *Nature* 241 (February 1973): 477–478. Reproduced with
permission from William C. McGrew.

appear to teach this technique to their offspring, flossing with more pauses, repeated motions, and for a longer time when their infants are nearby.

Macaques aren't the only flossing monkeys; a captive female baboon was photographed removing a bristle from a broom in her enclosure to floss her upper and lower teeth. Wild orangutans, common chimpanzees, and bonobo chimpanzees may also clean their teeth with twigs or sticks. Our young chimpanzee patient Bandit once used a ribbon-like piece of cloth to "floss" a loose baby tooth that he ultimately dislodged. Other captive chimpanzees presented with mirrors may sometimes use them to inspect and clean their teeth. While these actions may not leave indelible marks like the toothpick experiment, they firmly establish that nonhuman primates have an implicit awareness of their oral environment and the capacity to manipulate it.

The anecdotes I've shared might give you the impression that dental grooming is common in nonhuman primates, yet these behaviors are not seen very often. One innovative individual, or even population of individuals, may routinely care for their teeth, while other individuals or groups rarely copy their actions. This limits the exchange or transmission of these behaviors among groups or across generations.

Several ideas have been proposed to explain the more widespread adoption of toothpick use in the genus *Homo*.[40] Some suggest that gingival inflammation—or the periodontal disease it can lead to—drove hominins to use toothpicks. Ancient teeth with toothpick grooves often show a loosening of bony attachments, although it's important to bear in mind that fossilization can blur the picture of an individual's health prior to death. Peter Ungar and colleagues support the idea that toothpick use may be due to routine meat eating. It is difficult to assess whether eating tough or fibrous wild game motivated our ancestors to use toothpicks—if this was so, it seems unclear why toothpicking would have become commonplace only after 1.8 million years ago. In the preceding chapters we've discussed how evidence for animal butchery predates the earliest members of *Homo* by nearly a million years. Moreover, we aren't the only primates that eat meat. Certain chimpanzee communities routinely hunt monkeys, and orangutans occasionally catch and consume slow-moving primates called lorises.

Evidence from nonhuman primates and ancient hominins complicates the identification of any straightforward relationships among toothpick use, oral health, and diet. This leads me to wonder whether modern people who indulge in carnivorous diets seek out toothpicks more often than folks who enjoy plant-based or vegetarian diets? I relish the chance to pepper my dentist with questions like this when I head in for my semi-annual tooth cleaning!

Human behavior is notoriously difficult to study prior to recorded history. Because dental hard tissues preserve more frequently than other tissues of the body, they represent one of the most available sources of information about the differentiation of people and their cultures before written records. Dental practices like avulsion, filing, and inlay are direct evidence of an individual's behavior in the past. Other forms of body modification, including tattooing or scarification, aren't preserved for very long after death, and adornments like clothing are equally perishable. Deciphering the cultural messages of modified teeth is a challenging process. Burial and fossilization processes may obscure, distort, or destroy key evidence. Studies of recent small-scale societies have demonstrated that—as with many aspects of human behavior—place and time are essential considerations for the interpretation of this evocative form of symbolism. We do best to apply an anthropological framework that integrates knowledge of the biology of teeth, archeological science, nonhuman primate behavior, and cross-cultural perspectives on our humanity.

Concluding Thoughts: The Future That Teeth Foretell

Tiny records of cellular labor during tooth development permit an unparalleled glimpse into ancient humans' childhoods. Yet grappling with the fine details of tooth growth provides insights into much more than our past. Geneticists are working to engineer our future from the cellular dance that begins before we are born. Oral biologists are busy uncovering the environmental influences that lead to dental decay, misalignment, and tooth loss. These advances are particularly timely, since dental development in human populations may be accelerating, which could lead to greater incidences of these problems. More research is needed to probe changes in development over the past century, particularly to distinguish the influence of genetic and environmental factors.

There is no question that human biology continues to evolve—driven in part by the rapid cultural innovations that have allowed us to spread around the globe and swell rapidly in number. Fundamental changes during the Agricultural and Industrial Revolutions have had a profound impact on the development, evolution, and behavior of nearly all humans living today. Our dentitions continue to mark modern behaviors in myriad ways, serving as time capsules for future studies of our curious species. We can rest assured that whether or not teeth remain essential for noshing our highly processed diets, humanity will continue its fascination with them—adorning, collecting, and reviling them when they demand painful care.

Development

Tooth loss is a nearly universal human experience and a serious public health problem. In chapter 3, we discussed the archaic practice of implanting teeth from indigent human donors into the mouths of those who could afford them, a largely ineffective treatment. Fortunately, modern medicine is working to change this, and *bioengineering* is now a major endeavor for diverse research groups. The aim is to use cells from the patient's own body to grow replacement tissues or organs. This may ease the unmet demand for donated organs and lessen the body's rejection of foreign cells. Innovative mechanisms for creating replacement teeth may eventually take the place of dentures, implants, or bridges—expensive appliances that lack the vitality or sensory capabilities of healthy teeth. I've been following the field of restorative dentistry for the past decade, as it's been turbo-charged by breakthroughs in molecular biology, tissue engineering, and 3D printing technologies.[1]

You've likely heard or read about stem cell research—a subject often in the news due to controversy surrounding embryological tissues and human cloning. Stem cells have the remarkable ability to turn into any type of mature cell in the body depending on the chemical signals they receive and the genes that are consequently activated. This ability makes them particularly interesting to biologists—especially those who are working to repair or grow new tissues and organs. Within a few days after fertilization, human embryos form a small mass of these unspecialized cells, which can either recreate themselves over and over again, or take on a specific cellular fate. It turns out that adults carry similar cells in their bodies as well. Bone marrow, adipose tissue, and teeth each contain various types of stem cells that help the body rejuvenate our aging systems. Stem-cell therapies have been used to treat blood and immune disorders for years, and they are currently being tested to attenuate additional diseases and injuries.[2]

Research on dental stem cells has capitalized on the fact that certain mammals, including rodents and elephants, have continuously growing incisors.[3] The base of these teeth lies deep inside the jawbone and contains a region called a cervical loop. This is where epithelial and mesenchymal stem cells are directed to become enamel- and dentine-secreting cells, respectively. Once activated, these cells move out to secrete their hard tissues, and they must be replaced continuously to ensure the slow and steady growth of the incisors. Other animals such as zebrafish, which replace teeth throughout their lives, provide important insight into how stem cells enable this progressive development. Clinical researchers conduct experiments on these "model organisms," since much of the recipe for human tooth development has been retained over the course of vertebrate evolution.

Stem cells are also present in teeth that grow for a limited period of time, including the human dentition.[4] Recall that once tooth formation is complete, dentine-forming cells become trapped deep inside the tooth roots, where they slowly continue to secrete protective dentine into the pulp space. Small numbers of mesenchymal stem cells are also on hand, to be stimulated when heavy wear exposes the pulp and destroys mature dentine-forming cells. By remaining inside the root, these stem cells may replace the lost cells that repair dentine. Recent studies of mice have shown that it is possible to chemically nudge these cells, putting them to work when dentine is artificially damaged.[5] Future dentists may one day be able to use knowledge of molecular signaling pathways to turn on this patching process, which could lessen the need to create large dental fillings.

Importantly, stem cells can be harvested from the pulp tissue of human deciduous teeth that are naturally shed, as well as those that are clinically removed.[6] These cells can then be kept alive outside the body and stored for later use. There are even companies that "bank" your tooth stem cells for a fee—maintaining them for prospective future needs! One application being tested is the use of stems cells to regrow the tissue lost during a root canal procedure. When teeth are infected, current clinical practice involves removing the soft pulp, cleaning, and permanently filling the canal with dental cement. While this may prevent future infections, the tooth is no longer able to protect itself by producing reparative dentine. I'm looking forward to learning whether these clinical trials succeed, because they could make the dreaded root canal a thing of the past.

Another prospective use of dental stems cells is for bioengineering teeth. Scientists are working to grow new teeth outside the body that can be implanted to replace damaged or missing teeth—a kind of "holy grail" of restorative dentistry. While initial results were promising, more research is needed to refine the use of adult stem cells for dental engineering.[7] Part of the trouble is that these cells often behave differently under lab conditions than they do inside the body, complicating efforts to grow "test-tube" teeth for human use. Some scholars are pursuing complementary approaches that do not involve stem cells. For example, a team led by the biologist Takashi Tsuji has successfully built on the tissue recombination studies discussed in chapter 1.[8] They created early-stage developing tooth buds by combining mouse dental epithelial cells with mesenchymal cells in the lab, then implanted them into the jaws of adult mice that had recently lost molar teeth. In 34 of 60 cases, the implanted cells developed into a tooth that erupted into the mouth, eventually reaching the chewing plane. These teeth were functionally equal to those formed by the normal pathway, except that they were smaller than other molars. Scientists are working on understanding the determinants of tooth size and cusp patterning to be able to customize this approach for eventual human tooth engineering. The potential benefits are enormous—including orthodontic refinement, trauma repair, or the replacement of diseased teeth—as long as recipients can be patient enough to wait several years until their new teeth grow up!

Another type of dental bioengineering involves the use of special scaffolds or artificial tooth buds to help direct tissue development.[9] Oral biologist Pamela Yelick assembled a team that has pioneered cutting-edge methods to bioengineer teeth in animals, including rats and pigs. In a recent study, they created artificial tooth buds that formed secretory cells and supporting blood vessels. What's even more remarkable is the fact that the buds began producing mineralized tissues. These significant achievements—built on centuries of meticulous studies of tooth development—herald an exciting future for restorative dentistry!

Evolution

In the beginning of this book, I promised to tackle the thorny question of whether humans are still evolving. This subject takes on particular importance when we consider bioengineering missing body parts or fixing ones that aren't working correctly, which may interrupt the process of natural selection. Evolution occurs when there is a natural change in the genetic structure of a population, leading to the "survival of the fittest." This kind of change isn't easy to measure, particularly in slow growing, large, highly interwoven populations—we're really not ideal study subjects!

An example of evolution that I often discuss with my students is eyesight impairment, which may have been fatal in ancient environments full of predators and pitfalls. Today, modern corrective lenses and laser surgery have nearly leveled the evolutionary playing field, meaning that people like me with less-than-perfect vision aren't strongly disadvantaged. Genetic variations that lead to poorer vision may be more common today, as individuals with treatable conditions survive and leave more offspring than they may have under less forgiving primitive conditions. But there also environmental influences at work here too, as rates of nearsightedness have skyrocketed over the past century.[10]

Scientists who investigate human evolutionary change prefer to target measurable characteristics that are strongly influenced by genes. Long-term studies of more than 1,000 Australian twins have firmly established that tooth size relates more to "nature" than to "nurture."[11] The beauty in this system is that if tooth size changes across generations, it is likely that the genetic structure of the population has changed as well, meaning that is has evolved in the formal sense. While this sounds easy enough to determine, population-level studies of our long-lived species are costly and difficult to administer. One powerful example comes from a health survey that began in Framingham, Massachusetts in 1948, which is now the longest-running multigenerational medical study.[12] It began as a study of heart health, enrolling more than 5,000 people who agreed to have medical exams every few years. When this group of people had children, they were added to the study. This second generation also includes more than 5,000 individuals, who have now had children that make up a third-generation cohort.

Researchers estimated genetic change through time by measuring the health, physical characteristics, and reproductive outcomes of these women and then comparing them to their children. I'll spare you the complex theory and mathematics needed to convert measurements to rates of evolutionary change, focusing on the relevant results instead. The daughters of the original Framingham mothers are predicted to have their first child at slightly younger ages and reach menopause at slightly later ages—and the third generation is likely to continue this trend. This may surprise you, as many women in industrialized nations are having children at older ages than their mothers, or not having them at all. The case of the Framingham daughters shows a different trend, implying an evolutionary elongation of their "biological clocks."

This isn't the only study that finds natural selection is working to increase the reproductive period in women.[13] Evidence from Australian twins also shows that age at first reproduction is getting younger in this country as well. Similarly, women in a small French-Canadian village showed decreases in this age across generations, occurring prior to industrial development. You may be familiar with another oft-discussed trend in reproduction: girls in Western industrialized nations are now beginning their menstrual periods at younger ages than women in previous generations.[14] Scientists ascribe part of this to environmental factors, such as better nutrition and reduced physical activity among contemporary girls than those in earlier generations. The body can redirect surplus energy to speed up reproductive maturation sooner than in earlier generations. Similar developmental changes are seen across generations of well-nourished animals in captivity, as they are largely relieved of the energetic demands of foraging or hunting to meet their nutritional needs.

How do we parse out whether biological changes are due to environmental influences rather than genetic disposition? A study of Danish twins born before and after the two World Wars examined the determinants of reproduction.[15] The research team compared genetically similar individuals in different developmental environments, and unrelated individuals during more stable periods of history. They concluded that there is a strong genetic influence on a woman's ability to have children, and the number of children a woman bears is also influenced by social and economic pressures. In other words, fertility runs in families, but culture matters too. This leads us to search for other evidence of evolution that might be less affected by the social norms of our complex societies.

Skeletal anatomy might seem like a good place to start, and in fact I've highlighted a number of dental changes that occurred during our prehistory. In chapter 3, we learned that third molars are absent more often in recent humans than they were in the past. In chapter 5, we discussed the reduction in the size of our teeth over the last 10,000 years. Unfortunately, these studies don't compare related individuals over successive generations, which makes it impossible to untangle potential environmental effects from genetic ones. Among the rare exceptions to this constraint are two studies that suggest tooth size trends may now be moving in the opposite direction, as cohorts of British and Chinese children have slightly larger teeth than their parents.[16] Since we know that tooth size is largely controlled by genes, increases in these kids' teeth point to the idea that teeth are continuing to evolve.

Scientists have attempted to get around the paucity of anatomical data on related individuals by examining whether human development has changed through time. For example, evidence from the growth of the collarbone, the latest-forming bone in the body, reveals an acceleration of development since the early twentieth century.[17] Similar trends have been documented in our upper- and lower-limb bones. You may wonder whether this could be due to better nutrition and a more sedentary lifestyle for children in Western countries—factors that are suspected to contribute to the earlier onset of menstruation in girls. Teeth provide a useful case to explore this more fully, since their development is thought to be more genetically regulated than other aspects of skeletal or reproductive development. Tooth calcification has been compared through time in populations of children from Europe, America, and China.[18] Most studies have found that teeth are calcifying at earlier ages than they did 20–50 years ago. And girls seem to be speeding up faster than boys. While this is consistent with other studies of development we've reviewed, they involve comparisons of children who are unrelated to one another.[19] To be really sure, we'll need to recruit and track cohorts of related individuals like those in the Framingham Heart Study. Are you game to sign up your kids, grandkids, and great-grandkids for a personal demonstration that humans continue to evolve?

Behavior

In these considerations of recent human evolution, a primary challenge for scientists is to rule out whether the changes we observe over time can be explained by nongenetic reasons, such as behavioral or environmental influences. It is well established that the widespread adoption of agriculture and the industrial production of food have profoundly affected our physiology and anatomy, including our teeth. This includes skyrocketing rates of cavities, impaction, and alignment problems. Without understanding these behavioral changes over the past 10,000 years, we might mistakenly conclude that our teeth have evolved to be more susceptible to cavities. In truth, a species of bacteria evolved to take advantage of changes in our dietary behavior.

Studies of the diverse bacteria in our mouths, known collectively as the oral microbiome, hold great promise for helping us link ancient behaviors—including the onset of agricultural subsistence—with our present biology. Scientists are exploring the prevalence and evolution of disease-causing bacteria, and what the oral microbiome can tell us about prehistoric diets.[20] We discussed in chapter 7 how researchers armed with minute scrapings of tooth calculus have helped to identify the origins of animal milk consumption. The advent and spread of dairying over the last 8,000–9,000 years is a classic example of recent human evolution. Populations in regions with ancient dairying traditions show high frequencies of a gene that produces lactase, an enzyme that allows adults to digest milk. Prior to the domestication of dairy livestock, most humans lost this ability after infancy, as is the case for many Asian and African people today. Studies such as these are just getting started, since we have only recently gained the capability to probe the ancient DNA of bacteria and their human hosts, in addition to dietary proteins and microfossils, from tooth calculus.

Another type of cutting-edge science is piquing the curiosity of biological anthropologists keen to understand human behavior in the past.[21] During enamel formation, tiny amounts of protein become sandwiched between mineral crystallites, which stabilize the protein and preserve its original organic structure far longer than DNA typically survives in teeth. These proteins can be recovered from enamel and decoded into sequences of amino acids, some of which differ between females and males. The potential for determining the sex of humans from archeological sites is currently under investigation, and the initial results are exciting! As we've seen in chapter 8, standard approaches to determine the sex of individuals from their teeth are problematic, limiting our ability to understand ancient social relations— including aggressive interactions, territoriality, and reproductive dispersal. Scientists of the future may be able to subject a tooth to "nano-scale analysis" to learn more precisely who someone was, where they came from, what they ate, and how they relate to their group mates and the hominins that came before them.

Fine-scaled sampling also allows us to relate chemical changes to the daily records of tooth formation, so we can document the timing of behaviors such as breastfeeding and weaning prior to recorded history. My colleagues and I have done just that measuring barium intake, for example. The related field of environmental exposure biology is capitalizing on the identification of similar "biomarkers" in teeth, including lead and manganese. Ingestion of these metals has been linked to health problems in later life, including cognitive impairment.[22] The recent lead contamination of drinking water in Flint, Michigan, is likely to leave its signature in the teeth of the unfortunate children who drank local tap water during this period. Hopefully this type of information can lead to early interventions for those most at risk, although it will take years before the effects of this tragedy are fully understood.

An even darker tale told by teeth concerns environmental exposure to nuclear radiation.[23] Following the use of the atomic bomb during World War II, a group of politically active scientists, including Nobel Prize winner

Linus Pauling, began research with the aim to educate the public about the danger posed by nuclear weapons testing. This group launched the "Baby Tooth Survey" to assess the amount of a radioactive strontium isotope in children born in St. Louis, Missouri before, during, and after a period of especially heavy nuclear testing in 1953. The atomic variant of strontium they studied—strontium-90—behaves similarly to calcium, becoming concentrated in milk and captured in developing teeth and bone mineral. By collecting and analyzing 300,000 baby teeth, scientists were able to show that children who grew their teeth prior to 1953 had eight times less radioactive strontium than those born in 1957. All told, values of strontium-90, a cancer-causing agent, increased by nearly 50 times over 13 years in the children studied. These findings were presented to the U.S. Congress in 1963, which appear to have been persuasive. Later that year an international treaty was signed that partially banned the testing of nuclear weapons. One can't help but think of residents near nuclear disasters in Chernobyl, Ukraine and Fukushima, Japan. I dearly hope children born in the years to come won't bear any nuclear signatures in their teeth.

A final question I am often asked is, "What will future humans look like?" Presuming that human diets continue to center around soft, sugary, processed foods, our teeth won't wear down very fast, but they won't fit together very well either. We'll continue to need health care to combat rampant cavities and impacted molars, which may find even less space for eruption than they once had. Given our small, narrow jaws and large, lightly worn teeth, it's likely that the perfect smile will elude many of us. If recent trends in cosmetic dentistry are to be taken seriously, the resolution of this dilemma will depend on which culture or country one resides in. Some of us will certainly continue to express our individuality though our teeth, perhaps intentionally offsetting them, capping them with precious metals, or solemnly filing them down. You can be certain that the anthropologists of the future will have new toothy tales to ponder.

Acknowledgments

Janice Audet of Harvard University Press sparked this book, gently talking me through the gut-level anxiety of considering a post-PhD project of unknown depth and breadth. I am grateful for her interest, as well as that of Reed Malcolm at the University of California Press. Robert Prior, my editor at the MIT Press, was the picture of patience and goodwill as this book came together on two coasts, and was ultimately completed on two continents. His team at the Press, especially Katherine Almeida, Anne-Marie Bono, Christopher Eyer, and freelancer Kristie Reilly, have been responsive and professional.

I enlisted an invaluable cadre of friends and colleagues while researching and writing, including Manish Arora, Christine Austin, Lulu Cook, Christopher Dean, Arthur Durband, Mathieu Duval, Rebecca Ferrell, Rainer Grün, Philipp Gunz, Brian Hare, Leslea Hlusko, Zarin Machanda, Tesla Monson, Donald Reid, Paul Tafforeau, Nancy Tang, Mark Teaford, and Michael Westaway. Additional collaborators and colleagues at Stony Brook University, the Max Planck Institute, Harvard University, and Griffith University have profoundly expanded my understanding of human evolutionary biology over the past two decades. Kate Carter, William Delaney, Luca Fiorenza, Michael Foley, Michaela Huffman, Michelle Langley, Donald Reid, Holly Smith, and Katie Zink generously reviewed chapters. Any mistakes or omissions are mine.

Curating photographs and making figures was a true joy. The Peabody Museum of Archaeology and Ethnology at Harvard University allowed me to view human remains in their collections. I benefited from similar visits to the collections of the Phoebe A. Hearst Museum of Anthropology at the University of California at Berkeley while I was a visiting scholar there. Thank you to the custodians, including Michele Morgan and Olivia

Herschensohn at the Peabody Museum and Natasha Johnson and Benjamin Porter at the Hearst Museum. Acelyne Noël patiently captured my likeness in her first photo shoot. Those who provided original images or helped with permissions include Zeray Alemseged, Sakher AlQahtani, Bob Anemone, Christine Austin, Philippe Favier, Peggy Gough, Leslea Hlusko, Michaela Huffman, Véronique Huttman, Stella Ioannou, Aiko Kato, Reiko Kono, Erin Lawler, William McGrew, Michele Morgan, Masato Nakatsukasa, Matthew Skinner, Paul Tafforeau, Erik Trinkaus, and Tim White. I extend a deep bow of thanks for your help in bringing this subject alive.

My research has been funded by Stony Brook University, the Max Planck Society, Harvard University, the Radcliffe Institute for Advanced Study, and Griffith University, as well as the U.S. National Science Foundation, the Leakey Foundation, and the Wenner-Gren Foundation. Harvard University provided financial support for the preparation of this book, and assistance from Lenia Constantinou, Meg Lynch, and Monica Oyama. Griffith University helped me to get over the finish line, and I am particularly indebted to Rainer Grün and Dian Riseley for this.

This project was met with unwavering support from my friends and family, including Kate Amory, Shael Barger, Robin Feeney, Henry Kimsey-House, Zarin Machanda, Beth Russell, Rebecca Schlesinger, and my mother, Carol Smith. Bob Anemone, Daniel Green, and Don Reid buoyed me with their kindness and loyalty, as have my incredible student cohorts from HEB 1421 *Teeth*. Julie Silver and the 2015 participants of the Harvard Medical School Writing, Publishing, and Social Media for Healthcare Professionals program inspired and encouraged me. Matthew Battles, Elaine Davie, and Michael Fagan helped me to navigate the world of publishers and contracts. I also enjoyed an unforgettable writing residency in the company of several fantastic artists, photographers, and creative writers at PLAYA in Summer Lake, Oregon. Nina Katz and Craig Upson hosted me in their beautiful studio in Berkeley, Mary Smyer extended boundless compassion, and Leslea Hlusko and Tim White generously welcomed me into the Human Evolution Research Center at UC Berkeley. Janet Fink worked magic with words, healing energy, and her unflagging faith in me. A final word of thanks to Lulu Cook, who patiently continues to seed metta fields and joyful discourse each day in our home.

Notes

Introduction: Why Teeth?

1. Quoted on p. 1 of Simon Hillson, *Dental Anthropology* (Cambridge, UK: Cambridge University Press, 1996).

2. The term *dentine* was coined by the British anatomist Richard Owen but is often spelled "dentin" in the United States. I prefer "dentine" in homage to my British doctoral advisors and mentors, Lawrence Martin and Donald Reid. For more information see Michael John Trenouth, "The Origin of the Terms Enamel, Dentine and Cementum," *Faculty Dental Journal* 5 (January 2014): 27–31.

3. Michel Toussaint and Stéphane Pirson, "Neandertal Studies in Belgium: 2000–2005," *Periodicum Biologorum* 108 (2006): 373–387.

4. Teeth are estimated to make up 70% to 90% of all fossils recovered from several hominin sites in Africa, for example: Bernard A. Wood, "Tooth Size and Shape and Their Relevance to Studies of Hominid Evolution," *Philosophical Transactions of the Royal Society of London* B292 (1981): 65–76.

5. Zeresenay Alemseged et al., "A Juvenile Early Hominin Skeleton from Dikika, Ethiopia," *Nature* 443 (2006): 296–301.

6. Raymond A. Dart, "*Australopithecus africanus*: The Ape-Man of South Africa," *Nature* 115, no. 2884 (February 1925): 195–199.

7. Quote from ibid., 196.

8. Noreen von Cramon-Taubadel, "Global Human Mandibular Variation Reflects Differences in Agricultural and Hunter-Gatherer Subsistence Strategies," *Proceedings of the National Academy of Sciences USA* 108 (December 2011): 19546–19551.

Chapter 1

1. Described further in Alan Boyde, *The Structure and Development of Mammalian Enamel* (London: PhD dissertation, The London Hospital Medical College, 1964); William J. Croft, *Under the Microscope: A Brief History of Microscopy* (Singapore: World Scientific Press, 2006).

2. Quote from page 181 of Robert Hooke, *Micrographia* (London, Royal Society, 1665).

3. Dianna K. Padilla, "Inducible Phenotypic Plasticity of the Radula in *Lacuna* (Gastropoda: Littorinidae)," *The Veliger* 41 (April 1998): 201–204; Asa H. Barber, Dun Lu, and Nicola M. Pugno, "Extreme Strength Observed in Limpet Teeth," *Interface* 12 (2015): 20141326.

4. Quote from p. 1002 of Anthony Leeuwenhoeck, "Microscopical Observations of the Structure of Teeth and Other Bones: Made and Communicated in a Letter by Mr. Anthony Leeuwenhoeck," *Philosophical Transactions of the Royal Society of London* 12 (1665–1678): 1002–1003.

5. Technically, growth is an increase in size or number, while development is a more comprehensive term describing a transformation. Growth is encompassed within development.

6. The following sources provide more detailed reviews of the embryology and molecular biology of tooth formation: A. Richard Ten Cate, Paul T. Sharpe, Stéphane Roy, and Antonio Nancy, "Development of the Teeth and Its Supporting Structures," in *Ten Cate's Oral Histology*, ed. Antonio Nanci (St. Louis, MO: Mosby, 2003), 79–110; Andrew H. Jheon, Kerstin Seidel, Brian Biehs, and Ophir D. Klein, "From Molecules to Mastication: The Development and Evolution of Teeth," *WIREs Developmental Biology* (2012), doi: 10.1002/wdev.63; P. David Polly, "Gene Networks, Occlusal Clocks, and Functional Patches: New Understanding of Pattern and Process in the Evolution of the Dentition," *Odontology* 103 (2015): 117–125.

7. In contrast, bone is constantly being reworked, which removes and overwrites earlier records of its growth.

8. Described further in Peter Lucas, Paul Constantino, Bernard Wood, and Brian Lawn, "Dental Enamel as a Dietary Indicator in Mammals," *BioEssays* 30 (2008): 374–385.

9. I reviewed this topic in greater detail in my 2004 PhD thesis, available here: http://paleoanthro.org/dissertations/download/. There is a large literature on annual rings in trees; see, for example: Claudio S. Lisi et al., "Tree-Ring Formation, Radial Increment Periodicity, and Phenology of Tree Species From a Seasonal Semi-Deciduous Forest in Southeast Brazil," *IAWA Journal* 29 (2008): 189–207.

10. Evidence for daily and long-period rhythms in bone can be found in Hisashi Shinoda and Masahiro Okada, "Diurnal Rhythms in the Formation of Lamellar Bone in Young Growing Animals," *Proceedings of the Japanese Academy Series B* 64 (1988): 307–310; Timothy G. Bromage et al., "Lamellar Bone Is an Incremental Tissue Reconciling Enamel Rhythms, Body Size, and Organismal Life History," *Calcified Tissue International* (2009) 84: 388–404.

11. A curious historical detail about these studies is the fact that Okada submitted an English-language summary of his work to *The Shanghai Evening Post*, an American-owned newspaper originally published in China until the Japanese shut it down during World War II. The paper restarted again in New York in 1943, the year Okada published his summary—and one can only wonder why a Japanese scientist would report his research on tooth growth in a special edition of an anti-Japanese newspaper 2 years before the war ended: Masahiro Okada, "Hard Tissues of Animal Body: Highly Interesting Details of Nippon Studies in Periodic Patterns of Hard Tissues are Described," *The Shanghai Evening Post* (1943): 15–31. These studies are dramatized in Masahiro Okada, *The Hard Tissue: A Faithful Record of Metabolic Fluctuations* (video presented at the International Physiological Association Meetings, Leiden, 1963). These difficult to obtain sources are also described in M. C. Dean, "The Nature and Periodicity of Incremental Lines in Primate Dentine and Their Relationship to Periradicular Bands in OH 16 (*Homo habilis*)," in *Aspects of Dental Biology: Paleontology, Anthropology and Evolution*, ed. Jacopo Moggi-Cecchi (Florence: International Institute for the Study of Man, 1995), 239–265. Dean also reviews similar research from the 1930s by Isaac Schour and colleagues in the United States that informed Okada and Mimura's experimental design. The former research team developed experimental labeling protocols to determine the speed of hard tissue growth in a terminally ill human child and in other mammals.

12. Their model was subsequently expanded upon in Alan Boyde, "Enamel," in *Handbook of Microscopic Anatomy: Teeth*, vol. V, part 6, eds. A. Oksche and L. Vollrath (Berlin: Springer-Verlag, 1989), 309–473; M. C. Dean, "The Nature and Periodicity of Incremental Lines in Primate Dentine"; Tanya M. Smith, "Experimental Determination of the Periodicity of Incremental Features in Enamel," *Journal of Anatomy* 208 (2006): 99–114.

13. Described further in M. C. Dean, "The Nature and Periodicity of Incremental Lines in Primate Dentine"; Mie Ohtsuka and Hisashi Shinoda, "Ontogeny of Circadian Dentinogenesis in the Rat Incisor," *Archives of Oral Biology* 40, no. 6 (1995): 481–485; Tanya M. Smith, "Experimental Determination of the Periodicity of Incremental Features in Enamel."

14. G. D. Rosenberg and D. J. Simmons, "Rhythmic Dentinogenesis in the Rabbit Incisor: Circadian, Ultradian, and Infradian Periods," *Calcified Tissue International* 32 (1980): 29–44; Mie Ohtsuka-Isoya, Haruhide Hayashi, and Hisashi Shinoda, "Effect of Suprachiasmatic Nucleus Lesion on Circadian Dentin Increment in Rats,"

American Journal of Physiology—Regulatory, Integrative and Comparative Physiology 280 (2001): R1364–R1370.

15. These are called Retzius lines in enamel and Andresen lines in dentine, eponymous terms that reflect the scientists who described them. They manifest on the tooth and root surface as perikymata and periradicular bands, respectively. See M. C. Dean, "The Nature and Periodicity of Incremental Lines in Primate Dentine" for a review of the origins of these terms.

16. Gysi, A. "Metabolism in Adult Enamel," *Dental Digest* 37 (1931): 661–668.

17. Tanya M. Smith, "Incremental Dental Development: Methods and Applications in Hominoid Evolutionary Studies," *Journal of Human Evolution* 54 (2008): 205–24; Timothy G. Bromage et al., "Lamellar Bone Is an Incremental Tissue."

18. Daniel E. Lieberman, "The Biological Basis for Seasonal Increments in Dental Cementum and Their Application to Archaeological Research," *Journal of Archaeological Science* 21 (1994): 525–539. Ursula Wittwer-Backofen, Jutta Gampe, and James W. Vaupel, "Tooth Cementum Annulation for Age Estimation: Results From a Large Known-Age Validation Study," *American Journal of Physical Anthropology* 123 (2004): 119–129.

19. For example, see WeiJia Zhang, ZhengBin Li, and Yang Lei, "Experimental Measurement of Growth Patterns on Fossil Corals: Secular Variation in Ancient Earth-Sun Distances," *Chinese Science Bulletin* 55, no. 35 (2010): 4010–4017; M. Christopher Dean, "Progress in Understanding Hominoid Dental Development," *Journal of Anatomy* 197 (2000): 77–101; Tanya M. Smith, "Incremental Dental Development."

20. Described further in Alan Boyde, *The Structure and Development of Mammalian Enamel*; W. J. Schmidt and A. Keil, "Enamel," in *Polarizing Microscopy of Dental Tissues,* eds. W. J. Schmidt and A. Keil (Oxford: Pergamon Press, 1971), 319–427.

21. This is technically called propagation phase contrast X-ray synchrotron microtomography, and is described further in Paul Tafforeau and Tanya M. Smith, "Nondestructive Imaging of Hominoid Dental Microstructure Using Phase Contrast X-ray Synchrotron Microtomography," *Journal of Human Evolution* 54 (2008): 272–278. Also see Paul Tafforeau et al., "Applications of X-ray Synchrotron Microtomography for Nondestructive 3D Studies of Paleontological Specimens," *Applied Physics A* 83 (2006): 195–202.

22. Paul Tafforeau, John P. Zermeno, and Tanya M. Smith, "Tracking Cellular-level Enamel Growth and Structure in 4D with Synchrotron Imaging," *Journal of Human Evolution* 62, no. 3 (2012): 424–428.

Chapter 2

1. For more information, see Maury Massler, Isaac Schour, and Henry G. Poncher, "Developmental Pattern of the Child as Reflected in the Calcification Pattern of the Teeth," *American Journal of Diseases of Children* 629 (1941): 33–67; D. K. Whittaker and D. Richards, "Scanning Electron Microscopy of the Neonatal Line in Human Enamel," *Archives of Oral Biology* 23 (1978): 45–50; Jörgen G. Norén, "Microscopic Study of Enamel Defects in Deciduous Teeth of Infants Born to Diabetic Mothers," *Acta Odontologica Scandinavica* 42 (1984): 154–156; Wendy Birch and M. Christopher Dean, "Rates of Enamel Formation in Human Deciduous Teeth," in *Comparative Dental Morphology. Frontiers of Oral Biology*, vol. 13. (Basel: Karger, 2009), 116–120.

2. Jörgen G. Norén, "Microscopic Study of Enamel Defects in Deciduous Teeth of Infants Born to Diabetic Mothers"; but see Lotta Ranggård, Jörgen G. Norén, and Nina Nelson, "Clinical and Histologic Appearance in Enamel of Primary Teeth in Relation to Neonatal Blood Ionized Calcium Values," *Scandinavian Journal of Dental Research* 102 (1994): 254–259.

3. My colleagues and I have found similar disruptions during periods of weight loss in young monkeys long after birth, hinting at a causal connection among different types of accentuated lines in teeth. See Christine Austin et al., "Uncovering System-Specific Stress Signatures in Primate Teeth with Multimodal Imaging," *Science Reports* 5 (2016): 18802.

4. Ilana Eli, Haim Sarnat, and Eliezer Talmi, "Effect of the Birth Process on the Neonatal Line in Primary Tooth Enamel," *Pediatric Dentistry* 11 (1989): 220–223; Clément Zanolli, Luca Bondioli, Franz Manni, Paola Rossi, and Roberto Macchiarelli, "Gestation Length, Mode of Delivery and Neonatal Line Thickness Variation," *Human Biology* 83 (2011): 695–713. The authors of the second study did find that preterm babies had wider neonatal lines than those born on schedule and those born after their due date, although there was a fair degree of overlap in line widths among babies born after different gestation lengths.

5. At birth, teeth are small and conical, and—although teeth should be sliced perpendicular to the growth axis—it's easy to miss the tip of the growth center during sectioning. This could contribute to measurement error, as growth lines will be less well defined if there is even a small amount of angular deviation, and their thickness can be artificially exaggerated. This is an example of when a synchrotron imaging approach would be helpful for controlling the section plane and allowing more precise estimates of the thickness of the neonatal line.

6. Jeffrey H. Schwartz, Frank Houghton, Roberto Macchiarelli, and Luca Bondioli, "Skeletal Remains from Punic Carthage Do Not Support Systematic Sacrifice of Infants," *PLoS ONE* 5 (2012): e9177.

7. Wendy Birch and M. Christopher Dean, "A Method of Calculating Human Decid-uous Crown Formation Times and of Estimating the Chronological Ages of Stressful Events Occurring During Deciduous Enamel Formation," *Journal of Forensic and Legal Medicine* 22 (2014): 127–144.

8. For more information, see M. Christopher Dean and Fadil Elamin, "Parturition Lines in Modern Human Wisdom Tooth Roots: Do They Exist, Can They Be Char-acterized and Are They Useful for Retrospective Determination of Age at First Repro-duction and/or Inter-birth Intervals?" *Annals of Human Biology* 41 (2014): 358–367.

9. Jacqui E. Bowman, *Life History, Growth and Dental Development in Young Primates: A Study Using Captive Rhesus Macaques* (PhD dissertation, Cambridge University, Cam-bridge, 1991).

10. Christine Austin et al., "Uncovering System-Specific Stress Signatures in Primate Teeth with Multimodal Imaging."

11. B. R. Townend, "The Non-therapeutic Extraction of Teeth and Its Relation to the Ritual Disposal of Shed Deciduous Teeth," *British Dental Journal* 115, no. 8–10 (October 1963): 312–315, 354–357, 394–396; Rosemary Wells, "The Making of an Icon," in *The Good People: New Fairylore Essays*, ed. Peter Narváez (New York: Garland Publishing, 1991), 426–453; https://www.salon.com/2014/02/09/dont_tell_the_kids_the_real_history_of_the_tooth_fairy/; https://www.reuters.com/article/us-toothfairy-inflation-idUSBRE97T04M20130830/.

12. Deciduous tooth data from: L. Lysell, B. Magnusson, and Birgit Thilander, "Time and Order of Eruption of the Primary Teeth: A Longitudinal Study," *Odontologisk Revy* 13 (1962): 117–134. Permanent tooth data from Kaarina Haavikko, "The Formation and the Alveolar and Clinical Eruption of the Permanent Teeth: An Orthopantomographic Study," *Proceedings of the Finnish Dental Society* 66 (1970): 103–170; S. J. AlQahtani, M. P. Hector, and H. M. Liversidge, "Accuracy of Dental Age Estimation Charts: Schour and Massler, Ubelaker, and the London Atlas," *American Journal of Physical Anthropology* 154 (2014): 70–78. See this final source for ages of alveolar eruption and full eruption.

13. http://johnhawks.net/weblog/topics/history/aristotle_wisdom_teeth.html.

14. A. Richard Ten Cate and Antonio Nanci, "Physiologic Tooth Movement: Eruption and Shedding," in *Ten Cate's Oral Histology*, ed. Antonio Nanci (St. Louis, MO: Mosby, 2003), 275–298; M. Christopher Dean and Peter Vesey, "Preliminary Observations on Increasing Root Length During the Eruptive Phase of Tooth Development in Modern Humans and Great Apes," *Journal of Human Evolution* 54 (2008): 258–271.

15. A. Richard Ten Cate and Antonio Nanci, "Physiologic Tooth Movement: Erup-tion and Shedding."

16. For an overview of these different approaches, see Simon Hillson, *Dental Anthro-pology* (Cambridge, UK: Cambridge University Press, 1996).

17. S. J. AlQahtani, M. P. Hector, and H. M. Liversidge, "Accuracy of Dental Age Estimation Charts: Schour and Massler, Ubelaker, and the London Atlas," *American Journal of Physical Anthropology* 154 (2014): 70–78.

18. For more information, see Albert Edward William Miles, "Dentition in the Estimation of Age," *Journal of Dental Research* 42, suppl. to no. 1 (1963): 255–263; Richard Bassed, Jeremy Graham, and Jane A. Taylor, "Age Assessment," in *Forensic Odonotology: Principles and Practice*, eds. Jane A. Taylor and Jules A. Kieser (West Sussex, UK: Wiley Blackwell, 2016), 209–227; Edwin Saunders, *The Teeth a Test of Age, Considered with Reference to the Factory Children: Addressed to Members of Both Houses of Parliament* (London: H. Renshaw, 1837).

19. Ekta Priya, "Applicability of Willem's Method of Dental Age Assessment in 14 Years Threshold Children in South India—A Pilot Study," *Journal of Forensic Research* S4 (2015): S4–002.

20. Andreas Schmeling, Pedro Manuel Garamendi, Jose Luis Prieto, and María Irene Landa, "Forensic Age Estimation in Unaccompanied Minors and Young Living Adults," in *Forensic Medicine—From Old Problems to New Challenges*, ed. Duarte Nuno Vieira (Rijeka, Croatia: InTech Europe, 2011), 77–120.

21. This is provided that the individual has not completed root formation prior to death, since no additional days are added to the root after the root tip closes.

22. Because this repeat interval is constant within a tooth, it isn't necessary to count the days between each pair of long-period lines.

23. Pierre Formenty, "Ebola Virus Outbreak among Wild Chimpanzees Living in a Rain Forest of Côte d'Ivoire," *Journal of Infectious Diseases* 179, Suppl. 1 (1999): S120–S126.

24. The crown formation time for this cusp is 722 days, as it includes 31 days of prenatal enamel formation (691 + 31 days). Originally reported in Tanya M. Smith and Paul Tafforeau, "New Visions of Dental Tissue Research: Tooth Development, Chemistry, and Structure," *Evolutionary Anthropology* 17 (2008): 213–226.

25. For more information, see Daniel Antoine, Simon Hillson, and M. Christopher Dean, "The Developmental Clock of Dental Enamel: A Test for the Periodicity of Prism Cross-striations in Modern Humans and an Evaluation of the Most Likely Sources of Error in Histological Studies of this Kind," *Journal of Anatomy* 214 (2009): 45–55; Tanya M. Smith, "Teeth and Human Life-History Evolution," *Annual Review of Anthropology* 42 (2013):191–208.

26. Mark Skinner and Gail S. Anderson, "Individualization and Enamel Histology: A Case Report in Forensic Anthropology," *Journal of Forensic Sciences* 36 (1991): 939–948; M. A. Katzenberg et al., "Identification of Historical Human Skeletal Remains: A Case Study Using Skeletal and Dental Age, History and DNA," *International Journal of Osteoarcheology* 15 (2005): 61–72.

27. For more information, see Ursula Wittwer-Backofen et al., "Basics in Paleodemography: A Comparison of Age Indicators Applied to the Early Medieval Skeletal Sample of Lauchheim," *American Journal of Physical Anthropology* 137 (2008): 384–396; C. Cave and M. Oxenham, "Identification of the Archaeological 'Invisible Elderly': An Approach Illustrated with an Anglo-Saxon Example," *International Journal of Osteoarcheology* 26 (2016): 163–175.

28. J. A. Kieser, C. B. Preston, and W. G. Evans, "Skeletal Age at Death: An Evaluation of the Miles Method of Ageing," *Journal of Archaeological Science* 10 (1983): 9–12.

29. Discussed further in A. E. W. Miles, "Teeth as an Indicator of Age in Man," in *Development, Function, and Evolution of Teeth*, eds. P. M. Butler and K. A. Joysey (London, Academic Press, 1978), 455–464; Simon Hillson, *Dental Anthropology*.

30. Antonio Nanci, "Dentin-Pulp Complex," in *Ten Cate's Oral Histology*, ed. Antonio Nanci (Mosby St. Louis, MO, 2003), 192–239.

31. N. G. Clarke, S. E. Carey, W. Srikandi, R. S. Hirsch, and P. I. Leppard, "Periodontal Disease in Ancient Populations," *American Journal of Physical Anthropology* 71 (1986): 173–183.

32. Discussed further in P. Morris, "The Use of Teeth for Estimating the Age of Wild Mammals," in *Development, Function, and Evolution of Teeth*, eds. P. M. Butler and K. A. Joysey (London: Academic Press, 1978), 483–494; Richard F. Kay, D. Tab Rasmussen, and K. Christopher Beard, "Cementum Annulus Counts Provide a Means for Age Determination in *Macaca mulatta* (Primates, Anthropoidea)," *Folia Primatologica* 42 (1984): 85–95; Simon Hillson, *Dental Anthropology*; Ursula Wittwer-Backofen, Jutta Gampe, and James W. Vaupel, "Tooth Cementum Annulation for Age Estimation: Results From a Large Known-Age Validation Study," *American Journal of Physical Anthropology* 123 (2004): 119–129; H. Renz and R. J. Radlanski, "Incremental Lines in Root Cementum of Human Teeth a Reliable Age Marker?" *Homo* 57 (2006): 29–50; Stephan Naji et al., "Cementochronology, to Cut or Not to Cut?" *International Journal of Paleopathology* 15 (2016): 113–119.

33. Keith Condon, Douglas K. Charles, James M. Cheverud, and Jane E. Buikstra, "Cementum Annulation and Age Determination in *Homo sapiens*. II. Estimates and Accuracy," *American Journal of Physical Anthropology* 71 (1986): 321–330.

34. Ursula Wittwer-Backofen, Jutta Gampe, and James W. Vaupel, "Tooth Cementum Annulation for Age Estimation: Results from a Large Known-Age Validation Study."

35. Ursula Wittwer-Backofen et al., "Basics in Paleodemography."

36. Quote from p. 209 of Simon Hillson, *Dental Anthropology*.

37. Numerous presentations on cementum aging continue to be given at academic conferences, including special sessions at the 2012 and 2017 *American Association of Physical Anthropologists'* annual conferences, and more than a dozen presentations

at the *International Symposium on Dental Morphology* in 2017. See Stephan Naji et al., "Cementochronology, To Cut or Not to Cut?" for more context on this European-led resurgence in cementum annulation research.

Chapter 3

1. Daniel W. Sellen and Diana B. Smay, "Relationship Between Subsistence and Age at Weaning in 'Preindustrial' Societies." *Human Nature* 12, no. 1 (2001): 47–87.

2. Robert E. Black, Saul S. Morris, and Jennifer Bryce, "Where and Why Are 10 Million Children Dying Every Year?" *Lancet* 361 (June 2003): 2226–2234; Louise T. Humphrey, "Enamel Traces of Early Lifetime Events," in *Between Biology and Culture*, ed. Holger Schutkowski (Cambridge University Press, Cambridge, UK, 2008), 186–206.

3. Reidar F. Sognnaes, "Histological Evidence of Developmental Lesions in Teeth Originating From Paleolithic, Prehistoric, and Ancient Man," *The American Journal of Pathology* 32, no. 3 (1956): 547–577. Discussed further in Simon Hillson, *Tooth Development in Human Evolution and Bioarcheology* (Cambridge, UK: Cambridge University Press, 2014); Tanya M. Smith and Christopher Boesch, "Developmental Defects in the Teeth of Three Wild Chimpanzees from the Taï Forest," *American Journal of Physical Anthropology* 157 (2015): 556–570.

4. Wendy Birch and M. Christopher Dean, "A Method of Calculating Human Deciduous Crown Formation Times and of Estimating the Chronological Ages of Stressful Events Occurring During Deciduous Enamel Formation," *Journal of Forensic and Legal Medicine* 22 (2014): 127–144.

5. Examples are given from these studies: Jacqui E. Bowman, *Life History, Growth and Dental Development in Young Primates: A Study Using Captive Rhesus Macaques* (PhD dissertation, Cambridge University, Cambridge, 1991); Gary T. Schwartz, Don J. Reid, M. Christopher Dean, and Adrienne L. Zihlman, "A Faithful Record of Stressful Life Events Preserved in the Dental Developmental Record of a Juvenile Gorilla," *International Journal of Primatology* 27, no. 4 (August 2006): 1201–1219; Tanya M. Smith, "Teeth and Human Life-History Evolution," *Annual Review of Anthropology* 42 (2013): 191–208.

6. There may be at least 500 studies of enamel defects and nearly 100 identified causes, which are discussed further in Alan H. Goodman and Jerome C. Rose, "Dental Enamel Hypoplasias as Indicators of Nutritional Stress," in *Advances in Dental Anthropology*, eds. Marc A. Kelley and Clark Spencer Larsen (New York: Wiley-Liss, 1991), 279–293; Simon Hillson, *Tooth Development in Human Evolution and Bioarcheology*.

7. See illustrations in Stella Ioannou, Sadaf Sassani, Maciej Henneberg, and Renata J. Henneberg, "Diagnosing Congenital Syphilis Using Hutchinson's Method: Differentiating between Syphilitic, Mercurial, and Syphilitic-Mercurial Dental Defects," *American Journal of Physical Anthropology* 159 (2016): 617–629.

8. Jay Kelley, "Identification of a Single Birth Cohort in *Kenyapithecus kizili* and the Nature of Sympatry Between *K. kizili* and *Griphopithecus alpani* at Pasalar," *Journal of Human Evolution* 54 (2008): 530–537.

9. Quote from pp. 664–665 of Alfred Gysi, "Metabolism in Adult Enamel," *Dental Digest* 37 (1931): 661–668.

10. An example is provided from M. L. Blakey, T. E. Leslie, and J. P. Reidy, "Frequency and Chronological Distribution of Dental Enamel Hypoplasia," *American Journal of Physical Anthropology* 95 (1994): 371–383. This is discussed further in M. Katzenberg, D. A. Herring, and S. R. Saunders, "Weaning and Infant Mortality: Evaluating the Skeletal Evidence," *Yearbook of Physical Anthropology* 39 (1996): 177–99; Tanya M. Smith and Christopher Boesch, "Developmental Defects in the Teeth of Three Wild Chimpanzees from the Taï Forest."

11. Richard L. May, Alan H. Goodman, and Richard S. Meindl, "Response of Bone and Enamel Formation to Nutritional Supplementation and Morbidity Among Malnourished Guatemalan Children," *American Journal of Physical Anthropology* 92 (1993): 37–51. A broader consideration can be found in Alan H. Goodman and Jerome C. Rose, "Dental Enamel Hypoplasias as Indicators of Nutritional Stress."

12. Discussed further in Clark Spencer Larsen, "Biological Changes in Human Populations with Agriculture," *Annual Review of Anthropology* 24 (1995): 185–213; Simon Hillson, *Tooth Development in Human Evolution and Bioarcheology*; Daniel E. Lieberman, *The Story of the Human Body* (New York: Vintage Books, 2013).

13. Judith Littleton, "Invisible Impacts But Long-Term Consequences: Hypoplasia and Contact in Central Australia," *American Journal of Physical Anthropology* 126 (2005): 295–304.

14. See references in Clark Spencer Larsen, "Biological Changes in Human Populations with Agriculture."

15. Sergio Sergi, "Missing Teeth Inherited," *The Journal of Heredity* 5, no. 12 (1914): 559–560; K. Carter and S. Worthington, "Morphologic and Demographic Predictors of Third Molar Agenesis: A Systematic Review and Meta-Analysis," *Journal of Dental Research* 94, no. 7 (2015): 886–894; K. Carter and S. Worthington, "Predictors of Third Molar Impaction: A Systematic Review and Meta-analysis," *Journal of Dental Research* 95, no. 3 (2016): 267–276.

16. See references in Daniel E. Lieberman, *The Evolution of the Human Head* (Cambridge, MA: Harvard University Press, 2011); Katherine Carter, *The Evolution of Third Molar Agenesis and Impaction* (PhD dissertation, Harvard University, 2016). Evidence for M3 impaction in australopithecines is presented in Kathleen R. Gibson and James M. Calcagno, "Brief Communication: Possible Third Molar Impactions in the Hominid Fossil Record," *American Journal of Physical Anthropology* 91, no. 4 (August

1993): 517–521. However, it is difficult to know if these teeth might have been displaced after death.

17. Discussed further in Hannah J. O'Regan and Andrew C. Kitchener, "The Effects of Captivity on the Morphology of Captive, Domesticated and Feral Mammals," *Mammal Review* 35, nos. 3&4 (2005): 215–230; Daniel E. Lieberman, *The Evolution of the Human Head*.

18. Sergio Sergi, "Missing Teeth Inherited"; Daniel E. Lieberman, *The Evolution of the Human Head*.

19. James M. Calcagno and Kathleen R. Gibson, "Human Dental Reduction: Natural Selection or the Probable Mutation Effect," *American Journal of Physical Anthropology* 77 (1988): 505–517; Martin Kunkel, Wilfried Kleis, Thomas Morbach, and Wilfred Wagner, "Several Third Molar Complications Including Death—Lessons From 100 Cases Requiring Hospitalization," *Journal of Maxillofacial Surgery* 65 (2007): 1700–1706; C. Bowdler Henry and G. M. Morant, "A Preliminary Study of the Eruption of the Mandibular Third Molar Tooth in Man Based on Measurements Obtained from Radiographs, with Special Reference to the Problem of Predicting Cases of Ultimate Impaction of the Tooth," *Biometrika* 28, no. 3/4 (December 1936): 378–427; Mary Otte, *Teeth: The Story of Beauty, Inequality and the Struggle for Oral Health in America* (New York: The New Press, 2017).

20. In fairness to the bacteria that live in our mouths, many are beneficial for our health—or at least neutral in their impact—as is true of the bacteria in other parts of our digestive system.

21. Christina J. Adler et al., "Sequencing Ancient Calcified Dental Plaque Shows Changes in Oral Microbiota with Dietary Shifts of the Neolithic and Industrial Revolutions," *Nature Genetics* 45 (2013): 450–455; Christina Warinner et al., "Pathogens and Host Immunity in the Ancient Human Oral Cavity," *Nature Genetics* 46 (2014): 336–346.

22. For a general discussion of this topic, see Loren Cordain, *The Paleo Diet* (Boston: Houghton Mifflin Harcourt, 2011); Daniel E. Lieberman, *The Story of the Human Body*; Daniel H. Temple, "Caries: The Ancient Scourge," in *A Companion to Dental Anthropology*, eds. Joel D. Irish and G. Richard Scott (West Sussex, UK: John Wiley and Sons, 2016), 433–449; Luis Pezo Lanfranco and Sabine Eggers, "The Usefulness of Caries Frequency, Depth, and Location in Determining Cariogenicity and Past Subsistence: A Test on Early and Later Agriculturalists From the Peruvian Coast," *American Journal of Physical Anthropology* 143 (2010): 75–91. Specific studies discussed in the paragraphs that follow: Christina J. Adler et al., "Sequencing Ancient Calcified Dental Plaque Shows Changes in Oral Microbiota with Dietary Shifts of the Neolithic and Industrial Revolutions"; Omar E. Cornejo, "Evolutionary and Population Genomics of the Cavity Causing Bacteria *Streptococcus mutans*," *Molecular Biology and Evolution*

30, no. 4 (2013): 881–893; P.-F. Puech, H. Albertini, and N. T. W. Mills, "Dental Destruction in Broken-Hill Man," *Journal of Human Evolution* 9 (1980): 33–39.

23. Christy G. Turner II, "Dental Anthropological Indications of Agriculture Among the Jomon People of Central Japan," *American Journal of Physical Anthropology* 51 (1979): 619–636. Also see discussions in Clark Spencer Larsen, "Biological Changes in Human Populations with Agriculture"; Simon Hillson, "The Current State of Dental Decay," in *Technique and Application in Dental Anthropology*, eds. Joel D. Irish and Greg C. Nelson (Cambridge, UK: Cambridge University Press, 2008), 111–135.

24. These numbers represent conservative estimates of cavity prevalence, since it is often necessary to use radiography to locate cavities inside the tooth, which wasn't possible in Turner's study. Importantly, Turner omitted modern people consuming processed diets, as he felt that this would lead to artificially inflated rates. Given the differences in average values across subsistence groups, scientists have investigated whether the frequency of cavities can be used to predict the subsistence method of prehistoric humans. Unfortunately, this isn't as straightforward as one might hope. Populations within each subsistence type show considerable variation in the occurrence of cavities. For example, Turner's agricultural populations have frequencies that range from 2% in ancient Egyptians to 27% in the South Atlantic islanders from Tristan da Cunha. The lower end of the range for agriculturalists overlaps with hunter-gathers and populations that employed mixed-subsistence methods.

25. Nancy C. Lovell, *Patterns of Injury and Illness in Great Apes* (Washington, D.C.: Smithsonian Institution Press, 1990).

26. Louise T. Humphrey, "Earliest Evidence for Caries and Exploitation of Starchy Plant Foods in Pleistocene Hunter-Gatherers from Morocco," *Proceedings of the National Academy of Sciences USA* 111, no. 3 (January 2014): 954–959.

27. Simon Hillson, "The Current State of Dental Decay"; Ann Gibbons, "An Evolutionary Theory of Dentistry," *Science* 336 (May 2012): 973–975.

28. Elma Maria Vega Lizama and Andrea Cucina, "Maize Dependence or Market Integration? Caries Prevalence Among Indigenous Maya Communities With Maize-Based Versus Globalized Economies," *American Journal of Physical Anthropology* 153 (2014): 190–202. Of the 12 comparisons of specific age and sex groups between the two villages, one was tied at 100% of individuals affected in both communities.

29. John R. Lukacs and Leah L. Largaespada, "Explaining Sex Differences in Dental Caries Prevalence: Saliva, Hormones, and 'Life-History' Etiologies," *American Journal of Human Biology* 18 (2006): 540–555. Also see a critical discussion of this paper in Daniel H. Temple, "Caries: The Ancient Scourge."

30. See references in S. M. Hashim Nainar, "Is It Ethical to Withhold Restorative Dental Care From a Child with Occlusoproximal Caries Lesions Into Dentin of Primary Molars?" *Pediatric Dentistry* 37, no. 4 (July/August 2015): 329–331.

31. Bruce L. Pihlstrom, Bryan S. Michalowicz, and Newell W. Johnson, "Periodontal Diseases," *Lancet* 366 (2005): 1809–1820; Greg C. Nelson, "A Host of Other Dental Diseases and Disorders," in *A Companion to Dental Anthropology*, eds. Joel D. Irish and G. Richard Scott (West Sussex, UK: John Wiley and Sons, 2016), 465–483.

32. Bruce L. Pihlstrom, Bryan S. Michalowicz, and Newell W. Johnson, "Periodontal Diseases."

33. N. G. Clarke, S. E. Carey, W. Srikandi, R. S. Hirsch, and P. I. Leppard "Periodontal Disease in Ancient Populations," *American Journal of Physical Anthropology* 71 (1986): 173–183. Christina J. Adler et al., "Sequencing Ancient Calcified Dental Plaque Shows Changes in Oral Microbiota with Dietary Shifts of the Neolithic and Industrial Revolutions."

34. Ann Margvelashvili, Christoph P. E. Zollikofer, David Lordkipanidze, Paul Tafforeau, and Marcia S. Ponce de Leon, "Comparative Analysis of Dentognathic Pathologies in the Dmanisi Mandibles," *American Journal of Physical Anthropology* 160 (2016): 229–253; María Martinón-Torres et al., "Early Pleistocene Human Mandible from Sima del Elefante (TE) Cave Site in Sierra de Atapuerca (Spain): A Palaeopathological Study," *Journal of Human Evolution* 61 (2011): 1–11; Ana Gracia-Téllez et al., "Orofacial Pathology in *Homo heidelbergensis*: The case of Skull 5 from the Sima de los Huesos Site (Atapuerca, Spain)," *Quaternary International* 295 (2013): 83–93.

35. Stefan Baumgartner et al., "The Impact of the Stone Age Diet on Gingival Conditions in the Absence of Oral Hygiene," *Journal of Periodontology* 80, no. 5 (May 2009): 759–767.

36. James E. Anderson, "Human Skeletons of Tehuacán," *Science* 148, no. 3669 (April 1965): 496–497; Greg C. Nelson, "A Host of Other Dental Diseases and Disorders."

37. Lordkipanidze et al., "Anthropology: The Earliest Toothless Hominin Skull," *Nature* 434 (April 2005): 717–718.

38. Rachel Caspari and Sang-Hee Lee, "Older Age Becomes Common Late in Human Evolution," *Proceedings of the National Academy of Sciences USA* 101, no. 30 (July 2004): 10895–10900; Ann Margvelashvili, Christoph P. E. Zollikofer, David Lordkipanidze, Paul Tafforeau, and Marcia S. Ponce de Leon, "Comparative Analysis of Dentognathic Pathologies in the Dmanisi Mandibles."

39. S. Listl, J. Galloway, P. A. Mossey, and W. Marcenes, "Global Economic Impact of Dental Diseases," *Journal of Dental Research* 94, no. 10 (2015): 1355–1361.

40. Sydney Garfield, *Teeth Teeth Teeth* (New York: Simon and Schuster, 1969).

41. Robert S. Corrucini, "Anthropological Aspects of Orofacial and Occlusal Variations and Anomalies," in *Advances in Dental Anthropology*, eds. Marc A. Kelley and Clark Spencer Larson (New York: Wiley-Liss, 1991), 143–68; Jerome C. Rose and Richard D. Roblee, "Origins of Dental Crowding and Malocclusions: An Anthropological

Perspective," *Compendium of Continuing Education in Dentistry* 30, no. 5 (June 2009): 292–300.

42. P. R. Begg, "Stone Age Man's Dentition," *American Journal of Orthodontics* 40, no. 5 (1954): 373–383.

43. D. S. Carlson and D. P. Van Gerven, "Masticatory Function and Post-Pleistocene Evolution in Nubia," *American Journal of Physical Anthropology* 46, no. 3 (1977): 495–506.

44. Robert S. Corrucini, "Anthropological Aspects of Orofacial and Occlusal Variations and Anomalies."

45. Discussed further in Jerome C. Rose and Richard D. Roblee, "Origins of Dental Crowding and Malocclusions: An Anthropological Perspective"; M. Makaremi, K. Zink, and F. de Brondeau, "Apport des contraintes masticatrices fortes dans la stabilization de l'expansion maxillaire [The Importance of Elevated Masticatory Forces on the Stability of Maxillary Expansion]," *Revue d'Orthopédie Dento Faciale* 49 (2015): 11–20.

46. Robert S. Corruccini, *How Anthropology Informs the Orthodontic Diagnosis of Malocclusion's Causes* (Edwin Mellen Press, Lewiston, 1999); Daniel E. Lieberman, *The Evolution of the Human Head*; Ann Gibbons, "An Evolutionary Theory of Dentistry"; M. Makaremi, K. Zink, and F. de Brondeau, "Apport des contraintes masticatrices fortes dans la stabilisation de l'expansion maxillaire."

Chapter 4

1. Peter S. Ungar, *Teeth: A Very Short Introduction* (Oxford, UK: Oxford University Press, 2014).

2. See the following for different perspectives on this topic: Michal J. Benton, *Vertebrate Palaeontology* (West Sussex, UK: John Wiley and Sons, 2015); Neil Shubin, *Your Inner Fish: A Journey into the 3.5-Billion-Year History of the Human Body* (New York: Pantheon Books, 2009); Moya Meredith Smith et al., "Early Development of Rostrum Saw-Teeth in a Fossil Ray Tests Classical Theories of the Evolution of Vertebrate Dentitions," *Proceedings of the Royal Society of Biology Series B* 282 (2015): 20151628; Philip C. J. Donoghue and Martin Rücklin, "The Ins and Outs of the Evolutionary Origin of Teeth," *Evolution and Development* 18, no. 1 (2016): 19–30.

3. Donglei Chen, Henning Blom, Sophie Sanchez, Paul Tafforeau, and Per E. Ahlberg, "The Stem Osteichthyan *Andreolepis* and the Origin of Tooth Replacement," *Nature* 539 (2016): 237–224.

4. Jean-Yves Sire and Ann Huysseune, "Formation of Dermal Skeletal and Dental Tissues in Fish: A Comparative and Evolutionary Approach," *Biology Review* 78 (2003): 219–249; Qingming Qu, Tatjana Haitina, Min Zhu, and Per Erik Ahlberg,

"New Genomic and Fossil Data Illuminate the Origin of Enamel," *Nature* 526 (October 2015): 108–111; Martin D. Brazeau and Matt Friedman, "The Origin and Early Phylogenetic History of Jawed Vertebrates," *Nature* 520 (April 2015): 490–497.

5. Discussed further in Michal J. Benton, *Vertebrate Palaeontology*; Neil Shubin, *Your Inner Fish*.

6. Philip C. J. Donoghue and Martin Rücklin, "The Ins and Outs of the Evolutionary Origin of Teeth."

7. Images of shark and fish teeth can be seen in Barry K. Berkovitz and R. P. Shellis, *The Teeth of Non-Mammalian Vertebrates* (London: Academic Press, 2016).

8. Liam J. Rasch et al., "An Ancient Dental Gene Set Governs Development and Continuous Regeneration of Teeth in Sharks," *Developmental Biology* 415 (2016): 347–370; Donglei Chen, Henning Blom, Sophie Sanchez, Paul Tafforeau, and Per E. Ahlberg, "The Stem Osteichthyan *Andreolepis* and the Origin of Tooth Replacement."

9. Barry K. Berkovitz and R. P. Shellis, *The Teeth of Non-Mammalian Vertebrates*.

10. Discussed further in Michal J. Benton, *Vertebrate Palaeontology*; Jennifer A. Clack, "The Fish–Tetrapod Transition: New Fossils and Interpretations," *Evolution Education Outreach* 2 (2009): 213–223.

11. Edward B. Daeschler, Neil H. Shubin, and Farish A. Jenkins, "A Devonian Tetrapod-Like Fish and the Evolution of the Tetrapod Body Plan," *Nature* 440 (2006): 757–763; discussed further in Neil Shubin, *Your Inner Fish*.

12. Grzegorz Niedźwiedzki, Piotr Szrek, Katarzyna Narkiewicz, Marek Narkiewicz, and Per E. Ahlberg, "Tetrapod Trackways from the Early Middle Devonian Period of Poland," *Nature* 463 (January 2010): 43–48.

13. Another group of poorly studied legless amphibians known as cecilians show varied forms of reproduction, and do not rely as heavily on water as do other amphibians. Images of amphibian teeth can be found in Barry K. Berkovitz and R. P. Shellis, *The Teeth of Non-Mammalian Vertebrates*.

14. Discussed further in Michal J. Benton, *Vertebrate Palaeontology*; P. Martin Sander, "Reproduction in Early Amniotes," *Science* 337 (August 2012): 806–808.

15. Discussed further in Michal J. Benton, *Vertebrate Palaeontology*.

16. Discussed further in Michal J. Benton, *Vertebrate Palaeontology*; Matt Cartmill, William L. Hylander, and James Shafland, *Human Structure* (Cambridge, MA: Harvard University Press, 1987).

17. Paul C. Sereno, "Taxonomy, Morphology, Masticatory Function and Phylogeny of Heterodontosaurid Dinosaurs," *ZooKeys* 226 (2012): 1–225.

18. John M. Grady, Brian J. Enquist, Eva Dettweiler-Robinson, Natalie A. Wright, and Felisa A. Smith, "Evidence for Mesothermy in Dinosaurs," *Science* 344, no. 6189 (June 2014): 1268–1272; M. D. D'Emic, "Comment on "Evidence for Mesothermy in Dinosaurs" *Science* 348, no. 6238 (May 2015): 982; Robert A. Eagle et al., "Dinosaur Body Temperatures Determined from Isotopic (13C-18O) Ordering in Fossil Biominerals," *Science* 333, no. 6041 (June 2011): 443–445.

19. Pascal Godefroit et al., "A Jurassic Avialan Dinosaur from China Resolves the Early Phylogenetic History of Birds," *Nature* 498 (June 2013): 359–362.

20. E. J. Kollar and C. Fisher, "Tooth Induction in Chick Epithelium: Expression of Quiescent Genes for Enamel Synthesis," *Science* 207, no. 4434 (February 1980): 993–995.

21. Matthew P. Harris, Sean M. Hasso, Mark W. J. Ferguson, and John F. Fallon, "The Development of Archosaurian First-Generation Teeth in a Chicken Mutant," *Current Biology* 16 (February 2006): 371–377.

22. Recent work has argued that the genes needed to produce enamel have been irreparably altered in the chicken genome, meaning that if these mutants had survived they were unlikely to have been able to form fully functional teeth. Jean-Yves Sire, Sidney C. Delgado, and Marc Girondot, "Hen's Teeth with Enamel Cap: From Dream to Impossibility," *BMC Evolutionary Biology* 8, no. 246 (2008), doi: 10.1186/1471-2148-8-246.

23. Discussed further in Michael J. Benton, *Vertebrate Palaeontology*.

24. The technical term for having multiple tooth types is *heterodonty*, in contrast to *homodonty*, the condition of having a uniform tooth shape throughout the tooth row common to most fish, amphibians, reptiles, and dinosaurs. Discussed further in Z. Zhao, K. M. Weiss, and D. W. Stock, "Development and Evolution of Dentition Patterns and Their Genetic Basis," in *Development, Function and Evolution of Teeth*, eds. Mark F. Teaford, Moya Meredith Smith, and Mark W. J. Ferguson (Cambridge, UK: Cambridge University Press, 2007), 152–172.

25. B. Holly Smith, "'Schultz's Rule' and the Evolution of Tooth Emergence and Replacement Patterns in Primates and Ungulates," in *Development, Function and Evolution of Teeth*, eds. Mark F. Teaford, Moya Meredith Smith, and Mark W. J. Ferguson (Cambridge, UK: Cambridge University Press, 2007), 212–227.

26. Discussed further in Michael J. Benton, *Vertebrate Palaeontology*.

27. H. L. H. H. Green, "The Development and Morphology of the Teeth of *Ornithorhynchus*," *Philosophical Transactions of the Royal Society of London B* 288 (1937): 367–420; Masakazu Asahara, Masahiro Koizumi, Thomas E. Macrini, Suzanne J. Hand, and Michael Archer, "Comparative Cranial Morphology in Living and Extinct Platypuses: Feeding Behavior, Electroreception, and Loss of Teeth," *Science Advances* 2, no. 10 (2016): e1601329.

28. Barry Berkovitz, "Tooth Replacement Patterns in Non-Mammalian Vertebrates," in *Development, Function and Evolution of Teeth*, eds. Mark F. Teaford, Moya Meredith Smith, and Mark W. J. Ferguson (Cambridge, UK: Cambridge University Press, 2007), 186–200; Barry K. Berkovitz and R. P. Shellis, *The Teeth of Non-Mammalian Vertebrates*.

29. Alexander F. H. van Nievelt and Kathleen K. Smith, "To Replace or Not to Replace: The Significance of Reduced Functional Tooth Replacement in Marsupial and Placental Mammals," *Paleobiology*, 31, no. 2 (2005): 324–346.

30. See illustrations and discussion in Peter S. Ungar, *Mammal Teeth: Origin, Evolution, and Diversity* (Baltimore, MD: Johns Hopkins University Press, 2010); Simon Hillson, *Teeth* (Cambridge, UK: Cambridge University Press, 1986).

31. Oscar W. Johnson and Irven O. Buss, "Molariform Teeth of Male African Elephants in Relation to Age, Body Dimensions, and Growth," *Journal of Mammalogy* 46, no. 3 (August 1965): 373–384.

32. K. A. Kermack, Frances Mussett, and H. W. Rigney, "The Lower Jaw of *Morganucodon*," *Zoological Journal of the Linnean Society* 53 (September 1973): 87–175.

33. Discussed further in Peter S. Ungar, *Mammal Teeth*; Michael J. Benton, *Vertebrate Palaeontology*.

34. Discussed further in John G. Fleagle, *Primate Adaptation and Evolution* (San Diego, CA: Academic Press, 2013); Robert W. Sussman, D. Tab Rasmussen, and Peter H. Raven, "Rethinking Primate Origins Again," *American Journal of Primatology* 75 (2013): 95–106.

35. Xijun Ni, Qiang Li, Lüzhou Li, and K. Christopher Beard, "Oligocene Primates from China Reveal Divergence Between African and Asian Primate Evolution," *Science* 352 (2016): 673–677; Sunil Bajpai et al., "The Oldest Asian Record of Anthropoidea," *Proceedings of the National Academy of Sciences USA* 105 (2008): 11093–11098.

36. Dentists prefer to call them the first and second bicuspids.

37. Sunil Bajpai et al., "The Oldest Asian Record of Anthropoidea"; Xijun Ni, Qiang Li, Lüzhou Li, and K. Christopher Beard, "Oligocene Primates from China Reveal Divergence Between African and Asian Primate Evolution"; discussed further in John G. Fleagle, *Primate Adaptation and Evolution*.

38. Sally McBrearty and Nina G. Jablonski, "First Fossil Chimpanzee," *Nature* 437 (September 2005): 105–108; Brenda J. Bradley, "Reconstructing Phylogenies and Phenotypes: A Molecular View of Human Evolution," *Journal of Anatomy* 212 (2008) 337–353; Kevin E. Langergraber et al., "Generation Times in Wild Chimpanzees and Gorillas Suggest Earlier Divergence Times in Great Ape and Human Evolution," *Proceedings of the National Academy of Sciences USA* 109 (2102): 15716–15721.

Chapter 5

1. The Greek philosopher Aristotle actually produced a binomial classification system more than 2,000 years before Linnaeus, introducing the concepts of genus and species, as well as a logic-based comparative approach to classify organisms.

2. The species name *neanderthalensis* reflects the combination of "Neander" and "thal" (meaning "valley" in German); thus this name refers to the members of the genus *Homo* that originated from the Neander Valley. When the German language was modernized, the letter "h" was removed from the spelling of "thal," leading some to refer to this group as "Neandertals." While casual reference to either "Neanderthal" or "Neandertal" is accepted by paleoanthropologists, the formal name *Homo neanderthalensis* retains the original spelling, following the formal rules of taxonomic classification.

3. For more information, see John Reader, *Missing Links: In Search of Human Origins* (Oxford, UK: Oxford University Press, 2011); David Young, *The Discovery of Evolution* (Cambridge, UK: Cambridge University Press, 2007).

4. Dubois originally named the fossils from Java *Pithecanthropus erectus*, meaning "upright ape-like human," but the generic name *Pithecanthropus* was eventually disregarded since it was determined that this species was similar enough to other species in the genus *Homo* to be formally included. For more information, see Pat Shipman and Paul Storm, "Missing Links: Eugène Dubois and the Origins of Paleoanthropology," *Evolutionary Anthropology* 11 (2002): 108–116; John de Vos, "The Dubois Collection: A New Look at an Old Collection," *Scripta Geologic*, Special Issue 4 (2004): 267–285.

5. Kira E. Westaway et al., "An Early Modern Human Presence in Sumatra 73,000–63,000 Years Ago," *Nature* 548 (2017): 522–525. Also see the original description in D. A. Hoiijer, "Prehistoric Teeth of Man and of the Orang-utan from Central Sumatra, with Notes on the Fossil Orang-utan from Java and Southern China," *Zoologische Mededeelingen* 29 (1948): 175–301.

6. Pat Shipman and Paul Storm, "Missing Links: Eugène Dubois and the Origins of Paleoanthropology."

7. Many anthropologists consider that a fossil is a "hominin" when it has been shown to have any adaption for bipedal locomotion, irrespective of whether the majority of its skeletal features are ape-like. We will adopt this inclusive definition here for simplicity. When using "human" I refer only to *Homo sapiens*, as this is the only hominin species that possesses the full suite of traits that are found in living humans today.

8. Discussed further in Tanya M. Smith, Anthony J. Olejniczak, Stefan Reh, Donald J. Reid, and Jean-Jacques Hublin, "Brief Communication: Enamel Thickness Trends

in the Dental Arcade of Humans and Chimpanzees," *American Journal of Physical Anthropology* 136 (2008): 237–241.

9. Discussed further in Isabelle De Groote et al., "New Genetic and Morphological Evidence Suggests a Single Hoaxer Created 'Piltdown Man,'" *Royal Society Open Science* 3 (2016): 160328.

10. Gerrit S. Miller, "The Piltdown Jaw," *American Journal of Physical Anthropology* 1, no. 1 (1918): 25–52, with plates.

11. An interesting historical perspective is given by the founder of the *American Journal of Physical Anthropology* in the article preceding Miller's: Aleš Hrdlička, "Physical Anthropology: Its Scope and Aims; Its History and Present Status in America," *Journal of Physical Anthropology* 1, no. 1 (1918): 3–23.

12. Leonard Owen Greenfield, "Taxonomic Reassessment of Two *Ramapithecus* Species," *Folia Primatologica* 22 (1974): 97–115; David Pilbeam, "Hominoid Evolution and Hominoid Origins," *American Anthropologist* 88, no. 2 (1986): 295–312.

13. Quote from p. 35 of Gerrit S. Miller, "The Piltdown Jaw."

14. Tanya M. Smith et al., "Taxonomic Assessment of the Trinil Molars Using Non-Destructive 3D Structural and Development Analysis," *PaleoAnthropology* (2009): 117–129.

15. Tanya M. Smith, "Dental Development in Living and Fossil Orangutans," *Journal of Human Evolution* 94 (2016): 92–105.

16. Susana Carvalho, Eugenia Cunha, Cláudia Sousa, and Tetsuro Matsuzawa, "Chaînes Opératoires and Resource-Exploitation Strategies in Chimpanzee (*Pan troglodytes*) Nut Cracking," *Journal of Human Evolution* 55 (2008): 148–163; William C. McGrew, "In Search of the Last Common Ancestor: New Findings on Wild Chimpanzees," *Philosophical Transactions of the Royal Society B* 365 (2010): 3267–3276.

17. Tim D. White, Gen Suwa, and Berhane Asfaw, "*Australopithecus ramidus,* a New Species of Early Hominid from Aramis, Ethiopia," *Nature* 371 (1994): 306–312; Ann Gibbons, *The First Human* (New York: Doubleday, 2006).

18. Dates of first and last appearance taken from Bernard Wood and Eve K. Boyle, "Hominin Taxic Diversity: Fact or Fantasy?"

19. There is ongoing debate about whether all or any of these fossil taxa should be considered hominins; see, for example, Bernard Wood and Terry Harrison, "The Evolutionary Context of the First Hominins," *Nature* 470 (2011): 347–352.

20. Quote from p. 325 of Charles R. Darwin, *The Descent of Man, and Selection in Relation to Sex,* vol. 2 (London, UK: John Murray, 1871).

21. John Reader, *Missing Links: In Search of Human Origins.*

22. L. R. Berger and W. S. McGraw, "Further Evidence for Eagle Predation of, and Feeding Damage on, the Taung Child," *South African Journal of Science* 103 (2007): 496–498.

23. A report of a single robust australopithecine partial skeleton from east Africa suggests that they may have been fairly strong, muscular hominins. Manuel Domínguez-Rodrigo et al., "First Partial Skeleton of a 1.34-Million-Year-Old *Paranthropus boisei* from Bed II, Olduvai Gorge, Tanzania," *PLoS ONE* 8, no. 12 (2013): e80347.

24. Lee R. Berger et al., "*Australopithecus sediba*: A New Species of *Homo*-like Australopith from South Africa," *Science* 328 (2009): 195–204; Lee R. Berger, "The Mosaic Nature of *Australopithecus sediba*," *Science* 340 (2013): 163–165.

25. Shannon P. McPherron et al., "Evidence for Stone-Tool-Assisted Consumption of Animal Tissues Before 3.39 Million Years Ago at Dikika, Ethiopia," *Nature* 466 (2010): 857–860; Sonia Harmand et al., "3.3-Million-Year-Old Stone Tools from Lomekwi 3, West Turkana, Kenya," *Nature* 521 (2015): 310–315.

26. L. S. B. Leakey, P. V. Tobias, and J. R. Napier, "A New Species of the Genus *Homo* From the Olduvai Gorge," *Nature* 202 (1964): 7–9.

27. Bernard Wood and Mark Collard, "The Human Genus," *Science* 284 (1999): 65–71; Jeffrey H. Schwartz and Ian Tattersall, "Defining the Genus *Homo*," *Science* 349, no. 6251 (2015): 931–932.

28. Discussed further in Susan C. Antón, Richard Potts, and Leslie C. Aiello, "Evolution of Early *Homo*: An Integrated Biological Perspective," *Science* 345, no. 6192 (2014): 1236828. Climate cycles are illustrated in an online exhibit from the Smithsonian Institution: http://humanorigins.si.edu/evidence/human-evolution-timeline-interactive.

29. Brian Villmoare et al., "Early *Homo* at 2.8 Ma from Ledi-Geraru, Afar, Ethiopia," *Science* 347 no. 6228 (March 2015): 1352-1355; W. H. Kimbel et al., "Late Pliocene *Homo* and Oldowan Tools from the Hadar Formation (Kada Hadar Member), Ethiopia," *Journal of Human Evolution* 31 (1996): 549–561.

30. C. Loring Brace, "Environment, Tooth Form, and Size in the Pleistocene," *Journal of Dental Research* 46, suppl. to no. 5 (1967): 809–816.

31. John Hawks, Darryl J. de Ruiter, and Lee R. Berger, "Comment on "Early *Homo* at 2.8 Ma from Ledi-Geraru, Afar, Ethiopia," *Science* 348, no. 6241 (2015): 1326.

32. The classification of *Homo erectus* in Africa, Eurasia, and East Asia has been subject to debate, which is reviewed in Bernard Wood and Eve K. Boyle, "Hominin Taxic Diversity: Fact or Fantasy?" *Yearbook of Physical Anthropology* 159 (2016): S37–S78. I have used *Homo erectus* here to include all this material for simplicity.

33. David Lordkipanidze et al., "A Complete Skull from Dmanisi, Georgia, and the Evolutionary Biology of Early *Homo*," *Science* (2013) 342: 326–331. The primitive

nature of material from Georgia and Flores may imply that a species like *Homo habilis* left Africa, evolving into *Homo erectus* and then returning to Africa. Discussed further in Bernard Wood, "Did Early Homo Migrate 'Out Of' or 'In To' Africa?" *Proceedings of the National Academy of Sciences USA* 108 (2011): 10375–10376.

34. Dennis M. Bramble and Daniel E. Lieberman, "Endurance Running and the Evolution of *Homo*," *Nature* (2004) 432: 345–352; Katherine D. Zink and Daniel E. Lieberman, "Impact of Meat and Lower Palaeolithic Food Processing Techniques on Chewing in Humans," *Nature* (2016) 531: 500–503.

35. Anna-Sapfo Malaspinas et al., "A Genomic History of Aboriginal Australia," *Nature* 538 (2016): 207–214; Swapan Mallick et al., "The Simons Genome Diversity Project: 300 Genomes from 142 Diverse Populations," *Nature* 538 (2016): 201–206.

36. Matthias Meyer et al., "Nuclear DNA Sequences from the Middle Pleistocene Sima de los Huesos Hominins," *Nature* (2016) 531: 504–507.

37. Discussed further in Daniel E. Lieberman, "Speculations about the Selective Basis for Modern Human Craniofacial Form," *Evolutionary Anthropology* 17 (2008): 55–68; Chris Stringer, "The Origin and Evolution of *Homo sapiens*," *Philosophical Transactions of the Royal Society* B 371 (2016): 20150237.

38. Jean-Jacques Hublin et al., "New Fossils from Jebel Irhoud, Morocco and the Pan-African Origin of *Homo sapiens*," *Nature* 546 (2017): 289–292; Daniel Richter et al., "The Age of the Hominin Fossils from Jebel Irhoud, Morocco, and the Origins of the Middle Stone Age," *Nature* 546 (2017): 293–296.

39. Tanya M. Smith et al., "Dental Evidence for Ontogenetic Differences Between Modern Humans and Neanderthals," *Proceedings of the National Academy of Sciences USA* 107, no. 49 (2010): 20923–20928; Tanya M. Smith et al., "Variation in Enamel Thickness Within the Genus *Homo*," *Journal of Human Evolution* 62 (2012): 395–411.

40. C. Loring Brace, "Environment, Tooth Form, and Size in the Pleistocene"; Milford H. Wolpoff, *Metric Trends in Dental Evolution* (Cleveland, OH: Case Western Reserve University Press, 1971); Daniel E. Lieberman, *The Evolution of the Human Head* (Cambridge, MA: Harvard University Press, 2011); Debbie Guatelli-Steinberg, *What Teeth Reveal about Human Evolution* (Cambridge, UK: Cambridge University Press, 2016).

41. Albert A. Dahlberg, "Dental Evolution and Culture," *Human Biology* 35, no. 3 (1963): 237–249; C. Loring Brace, "Environment, Tooth Form, and Size in the Pleistocene"; Richard Wrangham, James Holland Jones, Greg Laden, David Pilbeam, and Nancy Lou Conklin-Brittain, "The Raw and the Stolen," *Current Anthropology* 40, no. 5 (1999): 567–593.

42. For an alternative perspective on this subject, see John A. J. Gowlett and Richard W. Wrangham, "Earliest Fire in Africa: Towards the Convergence of Archaeological

Evidence and the Cooking Hypothesis," *Azania: Archaeological Research in Africa*, 48, no. 1 (2013): 5–30.

43. Ron Shimelmitz et al., "'Fire at Will': The Emergence of Habitual Fire use 350,000 Years Ago," *Journal of Human Evolution* 77 (2014): 196–203.

44. Katherine D. Zink and Daniel E. Lieberman, "Impact of Meat and Lower Palaeolithic Food Processing Techniques on Chewing in Humans."

45. C. Loring Brace, "Environment, Tooth Form, and Size in the Pleistocene." See a discussion of the origins of this event in William L. Hylander and John T. Mayhall, "In Memoriam: Albert A. Dahlberg (1908-1993)," *American Journal of Physical Anthropology* 99 (1996): 627–633.

46. For example, see Patricia Smith, Yochanan Wax, Fanny Adler, Uri Silberman, and Gady Heinic, "Post-Pleistocene Changes in Tooth Root and Jaw Relationships," *American Journal of Physical Anthropology* (1986) 70: 339–348; James M. Calcagno, "Dental Reduction in Post-Pleistocene Nubia," *American Journal of Physical Anthropology* 70 (1986): 349–363.

47. Clark Spencer Larsen, "The Agricultural Revolution as Environmental Catastrophe: Implications for Health and Lifestyle in the Holocene," *Quaternary International* 150, no. 1 (2006): 12–20; Amanda Mummert, Emily Esche, Joshua Robinson, and George J. Armelagos, "Stature and Robusticity During the Agricultural Transition: Evidence from the Bioarchaeological Record," *Economics and Human Biology* 9, no. 3 (2011): 284–301.

48. C. Loring Brace, "Australian Tooth-Size Clines and the Death of a Stereotype," *Current Anthropology* (1980) 21: 141–164; C. Loring Brace, Karen R. Rosenberg, and Kevin D. Hunt, "Gradual Changes in Human Tooth Size in the Late Pleistocene and Post-Pleistocene," *Evolution* 41, no. 4 (1987): 705–720.

49. Reviewed in Debbie Guatelli-Steinberg, *What Teeth Reveal About Human Evolution*.

50. James M. Calcagno and Kathleen R. Gibson, "Human Dental Reduction: Natural Selection or the Probable Mutation Effect," *American Journal of Physical Anthropology* 77 (1988): 505–517.

51. See, for example, Terry Harrison, Changzhu Jin, Yingqi Zhang, Yuan Wang, and Min Zhu, "Fossil *Pongo* from the Early Pleistocene *Gigantopithecus* Fauna of Chongzuo, Guangxi, Southern China," *Quaternary International* 354 (2014): 59–67. Data on changes in the loss of enamel and dentine can be found in Tanya M. Smith et al., "Dental Tissue Proportions in Fossil Orangutans From Mainland Asia and Indonesia," *Human Origins Research* (2011): 1:e1.

52. Richard J. Smith and David R. Pilbeam, "Evolution of the Orang-utan," *Nature* 284 (1980): 447–448.

53. Thomas Sutikna et al., "Revised Stratigraphy and Chronology for *Homo floresiensis* at Liang Bua in Indonesia," *Nature* 532 (2016): 366–369; Gerrit D. van den Bergh et al., "*Homo floresiensis*-like Fossils From the Early Middle Pleistocene of Flores," *Nature* 534 (2016): 245–248; Debbie Argue, Colin P. Groves, Michael S. Y. Lee, and William L. Jungers, "The Affinities of *Homo floresiensis* Based on Phylogenetic Analyses of Cranial, Dental, and Postcranial Characters," *Journal of Human Evolution* 107 (2017): 107–133.

54. An interesting mechanistic explanation for tooth size reduction is the inhibitory cascade model, reviewed in Debbie Guatelli-Steinberg, *What Teeth Reveal About Human Evolution*. This model was recently applied to the human fossil record, suggesting that the reduction of tooth size in *Homo* is linked to a change in the proportions of the front and back molars: Alistair R. Evans et al., "A Simple Rule Governs the Evolution and Development of Hominin Tooth Size," *Nature* 530 (2016): 477–480.

Chapter 6

1. Data from Melissa Emery Thompson, "Comparative Reproductive Energetics of Human and Nonhuman Primates," *Annual Review of Anthropology* 42 (2013): 287–304; Melissa Emery Thompson, "Reproductive Ecology of Female Chimpanzees," *American Journal of Primatology* 75 (2013): 222–237; Fernando Colchero et al., "The Emergence of Longevous Populations," *Proceedings of the National Academy of Sciences USA* 113, no. 48 (2016): E7681–E7690; Shannen L. Robson and Bernard Wood, "Hominin Life History: Reconstruction and Evolution," *Journal of Anatomy* 212 (2008): 394–425; Daniel W. Sellen, "Comparison of Infant Feeding Patterns Reported for Nonindustrial Populations with Current Recommendations," *The Journal of Nutrition* 131 (2001): 2707–2715; Anne E. Pusey, "Mother-Offspring Relationships in Chimpanzees After Weaning," *Animal Behaviour* 31 (1983): 363–377; Ramon J. Rhine, Guy W. Norton, and Samuel K. Wasser, "Lifetime Reproductive Success, Longevity, and Reproductive Life History of Female Yellow Baboons (*Papio cynocephalus*) of Mikumi National Park, Tanzania," *American Journal of Primatology* 51, no. 4 (2000): 229–241.

2. B. Holly Smith, "Life History and the Evolution of Human Maturation," *Evolutionary Anthropology* 1 (1992): 134–42; Tanya M. Smith, "Teeth and Human Life-History Evolution," *Annual Review of Anthropology* 42 (2013): 191–208.

3. Melissa Emery Thompson, "Comparative Reproductive Energetics of Human and Nonhuman Primates,"; Fernando Colchero et al., "The Emergence of Longevous Populations."

4. Tanya M. Smith, Christine Austin, Katie Hinde, Erin R. Vogel, and Manish Arora, "Cyclical Nursing Patterns in Wild Orangutans," *Science Advances* 3 (2017): e1601517.

5. James J. McKenna, "The Evolution of Allomothering Behavior Among Colobine Monkeys: Function and Opportunism in Evolution," *American Anthropologist*

81 (1979): 818–840; Lynn A. Fairbanks, "Reciprocal Benefits of Allomothering for Female Vervet Monkeys," *Animal Behavior* 40 (1990): 553–562.

6. Melissa Emery Thompson, "Faster Reproductive Rates Trade Off Against Offspring Growth in Wild Chimpanzees," *Proceedings of the National Academies of Science USA* 113 (2016): 7780–7785.

7. K. Hawkes, J. F. O'Connell, N. G. Blurton Jones, H. Alvarez, and E. L. Charnov, "Grandmothering, Menopause, and the Evolution of Human Life Histories," *Proceedings of the National Academy of Sciences USA* (1998) 95: 1336–1339; Karen L. Kramer, "Cooperative Breeding and its Significance to the Demographic Success of Humans," *Annual Review of Anthropology* 39 (2010): 417–36.

8. Discussed further in Karen L. Kramer, Russell D. Greaves, and Peter T. Ellison, "Early Reproductive Maturity Among Pume Foragers: Implications of a Pooled Energy Model to Fast Life Histories," *American Journal of Human Biology* 21 (2009): 430–37.

9. Fernando Colchero et al., "The Emergence of Longevous Populations."

10. Kim Hill et al., "Mortality Rates Among Wild Chimpanzees," *Journal of Human Evolution* 40 (2001): 437–450.

11. For more information, see B. Holly Smith, Tracey L. Crummett, and Kari L. Brandt, "Ages of Eruption of Primate Teeth: a Compendium for Aging Individuals or Comparing Life Histories," *Yearbook of Physical Anthropology* 37 (1994): 177–231.

12. B. Holly Smith, "Dental Development as a Measure of Life History Variation in Primates," *Evolution* 43 (1989): 683–88.

13. Discussed further in Shannen L. Robson and Bernard Wood, "Hominin Life History: Reconstruction and Evolution," Louise T. Humphrey, "Weaning Behaviour in Human Evolution," *Seminars in Cell and Developmental Biology* 21 (2010): 453–461; Tanya M. Smith, "Teeth and Human Life-History Evolution."

14. Tanya M. Smith et al., "First Molar Eruption, Weaning, and Life History in Living Wild Chimpanzees,"; *Proceedings of the National Academy of Sciences USA* 110 (2013): 2787–2791; Zarin Machanda et al., "Dental Eruption in East African Wild Chimpanzees," *Journal of Human Evolution* 82 (2015): 137–144.

15. We're not the first to suggest that diet has a strong influence on primate molar eruption ages; these authors argued that this is true irrespective of life-history timing: Laurie R. Godfrey, K. E. Samonds, W. L. Jungers, and M. R. Sutherland, "Teeth, Brains, and Primate Life Histories," *American Journal of Physical Anthropology* 114 (2001): 192–214.

16. Discussed further in Melissa Emery Thompson, "Reproductive Ecology of Female Chimpanzees."

17. For example, see Aleš Hrdlička, "The Taungs Ape," *American Journal of Physical Anthropology* 8, no. 4 (1925): 379–392.

18. Donald Reid and M. Christopher Dean, "Variation in Modern Human Enamel Formation Times," *Journal of Human Evolution* 50 (2006): 329–346; Robert Walker et al., "Growth Rates and Life Histories in Twenty-Two Small-Scale Societies," *American Journal of Human Biology* 18 (2006): 295–311; Helen M. Liversidge, "Timing of Human Mandibular Third Molar Formation," *Annuals of Human Biology* 35 (2008): 294–321.

19. Timothy G. Bromage and M. Christopher Dean, "Re-Evaluation of the Age at Death of Immature Fossil Hominids," *Nature* 317 (1985): 525–527.

20. Rodrigo S. Lacruz, Fernando Ramirez Rozzi, and Timothy G. Bromage, "Dental Enamel Hypoplasia, Age at Death, and Weaning in the Taung Child," *South African Journal of Science* 101 (2005): 567–69. For information on the revised age of Sts 24, see Tanya M. Smith et al., "Dental Ontogeny in Pliocene and Early Pleistocene Hominins," *PLoS ONE* 10 (2015): e0118118

21. M. Christopher Dean et al., "Growth Processes in Teeth Distinguish Modern Humans from *Homo erectus* and Earlier Hominins," *Nature* 414 (2001): 628–631; José María Bermúdez de Castro et al.," New Immature Hominin Fossil from European Lower Pleistocene Shows the Earliest Evidence of a Modern Human Dental Development Pattern," *Proceedings of the National Academies of Science USA* 107 (2010): 11739–11744; Melvin Konner, *The Evolution of Childhood: Relationships, Emotions, Mind* (Cambridge, MA: Belknap Press, 2010).

22. Ronda R. Graves, Amy C. Lupo, Robert C. McCarthy, Daniel J. Wescott, and Deborah L. Cunningham, "Just How Strapping was KNM-WT 15000?" *Journal of Human Evolution* 59 (2010): 542–554.

23. Jeremy M. DeSilva and Julie J. Lesnik, "Brain Size at Birth Throughout Human Evolution: A New Method for Estimating Neonatal Brain Size in Hominins," *Journal of Human Evolution* 55 (2008): 1064–1074; M. Christopher Dean and B. Holly Smith, "Growth and Development of the Nariokotome Youth, KNM-WT 15000," in *The First Humans: Origin and Early Evolution of the Genus Homo*, eds. Frederick E. Grine, John G. Fleagle, and Richard E. Leakey (New York: Springer, 2009), 101–20; M. Christopher Dean, "Measures of Maturation in Early Fossil Hominins: Events at the First Transition from Australopiths to Early *Homo*," *Philosophical Transactions of the Royal Society B* 371 (2016): 20150234.

24. See extended discussions in Debbie Guatelli-Steinberg, "Recent Studies of Dental Development in Neandertals: Implications for Neandertal Life Histories," *Evolutionary Anthropology* 18 (2009): 9–20; Tanya M. Smith, "Teeth and Human Life-History Evolution."

25. Tanya M. Smith et al., "Earliest Evidence of Modern Human Life History in North African Early *Homo sapiens*," *Proceedings of the National Academies of Science*

USA 104 (2007): 6128–6133; Tanya M. Smith et al., "Dental Evidence for Ontogenetic Differences Between Modern Humans and Neanderthals." *Proceedings of the National Academies of Science USA* 107 (2010): 20923–20928.

26. At the time we published our developmental study, this fossil was estimated to be 160,000 years old, but the date was subsequently revised to approximately 300,000 years old, due in part to a mathematical error in the original calculation and the addition of newly dated sediments detailed in Daniel Richter et al., "The Age of the Hominin Fossils From Jebel Irhoud, Morocco, and the Origins of the Middle Stone Age," *Nature* 546 (2017): 293–296.

27. Michel Toussaint and Stéphane Pirson, "Neandertal Studies in Belgium: 2000–2005," *Periodicum Biologorum* 108, no. 3 (2006): 373–387.

28. See extended discussion in Jean-Jacques Hublin, Simon Neubauer, and Philipp Gunz, "Brain Ontogeny and Life History in Pleistocene Hominins," *Philosophical Transactions of the Royal Society* B 370 (2015): 20140062. Further information on dental differences can be found here: Clément Zanolli, Mathilde Hourset, Rémi Esclassan, and Catherine Mollereau, "Neanderthal and Denisova Tooth Protein Variants in Present-Day Humans," *PLoS ONE* 12 (2017): e0183802.

29. Holly M. Dunsworth, Anna G. Warrener, Terrence Deacon, Peter T. Ellison, and Herman Pontzer, "Metabolic Hypothesis for Human Altriciality," *Proceedings of the National Academies of Science USA* 109 (2012): 15212–15216.

30. B. Holly Smith, "Dental Development as a Measure of Life History Variation in Primates"; Steven R. Leigh and Gregory E. Blomquist, "Life History," in *Primates in Perspective*, eds. Christina J. Campbell, Agustin Fuentes, Katherine C. MacKinnon, Melissa Panger, and Simon K. Bearder (Oxford, UK: Oxford University Press, 2007), 396–407; Shannen L. Robson and Bernard Wood, "Hominin Life History: Reconstruction and Evolution"; Christopher W. Kuzawa et al., "Metabolic Costs and Evolutionary Implications of Human Brain Development," *Proceedings of the National Academies of Science USA* 111 (2014): 13010–13015.

31. B. Holly Smith, "Dental Development as a Measure of Life History Variation in Primates"; Barry Bogin, *Patterns of Human Growth* (Cambridge, UK: Cambridge University Press, 1999); Sue Taylor Parker, "Evolutionary Relationships Between Molar Eruption and Cognitive Development in Anthropoid Primates," in *Human Evolution Through Developmental Change*, eds. Nancy Minugh-Purvis and Kenneth J. McNamara (Baltimore, MD: The Johns Hopkins University Press, 2002), 305–316.

32. Richard J. Smith, Patrick J. Gannon, and B. Holly Smith, "Ontogeny of Australopithecines and Early *Homo*: Evidence from Cranial Capacity and Dental Eruption," *Journal of Human Evolution* 29 (1995): 155–68.

33. A recent example can be found in Antonio Rosas et al., "The Growth Pattern of Neandertals, Reconstructed from a Juvenile Skeleton from El Sidrón (Spain)," *Science* 357 (2017): 1282–1287.

34. Jay N. Giedd et al., "Brain Development During Childhood and Adolescence: A Longitudinal MRI Study," *Nature Neuroscience* 2 (1999): 861–863; Daniel E. Lieberman, *The Evolution of the Human Head* (Cambridge, MA: Harvard University Press, 2011); Jean-Jacques Hublin, Simon Neubauer, and Philipp Gunz, "Brain Ontogeny and Life History in Pleistocene Hominins."

35. Tomáš Paus, Matcheri Keshavan, and Jay N. Giedd, "Why Do Many Psychiatric Disorders Emerge During Adolescence?" *Nature Reviews* 9 (2008): 947–957; José María Bermúdez de Castro, Mario Modesto-Mata, and María Martinón-Torres, "Brains, Teeth and Life Histories in Hominins: A Review," *Journal of Anthropological Sciences* 93 (2015): 21–42.

36. Shannen L. Robson and Bernard Wood, "Hominin Life History: Reconstruction and Evolution"; Marcia S. Ponce de León et al., "Neanderthal Brain Size at Birth Provides Insights into the Evolution of Human Life History," *Proceedings of the National Academies of Science USA* 105: 13764–13768; but see Jay Kelley and Gary T. Schwartz, "Life-history Inference in the Early Hominins *Australopithecus* and *Paranthropus*," *International Journal of Primatology* 33 (2012): 1332–1363.

37. Steve R. Leigh, "Brain Growth, Life History, and Cognition in Primate and Human Evolution," *American Journal of Primatology* 62 (2004): 139–164; Jean-Jacques Hublin, Simon Neubauer, and Philipp Gunz, "Brain Ontogeny and Life History in Pleistocene Hominins."

38. See, for example, Christopher B. Stringer, M. Christopher Dean, and Robert D. Martin, "A Comparative Study of Cranial and Dental Development Within a Recent British Sample and Among Neandertals," in *Primate Life History and Evolution*, ed. C. J. De Rousseau (New York: Wiley-Liss, 1990), 115–152; Hélène Coqueugniot and Jean-Jacques Hublin, "Endocranial Volume and Brain Growth in Immature Neanderthals," *Periodicum Biologorum* 109 (2007): 379–385; Marcia S. Ponce de León, Thibaut Bienvenu, Takeru Akazawa, and Christoph P. E. Zollikofer, "Brain Development is Similar in Neanderthals and Modern Humans," *Current Biology* 26 (2016): R641–R666; Antonio Rosas et al., "The Growth Pattern of Neandertals, Reconstructed from a Juvenile Skeleton from El Sidrón (Spain)."

39. Rachel Caspari and Sang-Hee Lee, "Older Age Becomes Common Late in Human Evolution," *Proceedings of the National Academies of Science USA* 101 (2004): 10895–10900.

40. Fernando Colchero et al., "The Emergence of Longevous Populations."

41. Tanya M. Smith, "Teeth and Human Life-History Evolution."

42. See, for example, Hillard Kaplan, Kim Hill, Jane Lancaster, and A. Magdalena Hurtado, "A Theory of Human Life History Evolution: Diet, Intelligence, and Longevity," *Evolutionary Anthropology* 9 (2000): 156–85; Phyllis C. Lee, "Growth and Investment in Hominin Life History Evolution: Patterns, Processes, and Outcomes," *International Journal of Primatology* 33 (2012): 1309–1331; M. Christopher Dean, "Measures of Maturation in Early Fossil Hominins: Events at the First Transition from Australopiths to Early *Homo*."

43. Christopher W. Kuzawa et al., "Metabolic Costs and Evolutionary Implications of Human Brain Development"; Steve R. Leigh, "Brain Growth, Life History, and Cognition in Primate and Human Evolution."

44. But, not to be outdone, orangutan males may delay their final skeletal maturation for up to 10 years after sexual maturation, see Melissa Emery Thompson, Amy Zhou, and Cheryl D. Knott, "Low Testosterone Correlates with Delayed Development in Male Orangutans," *PLoS ONE* 7 (2012): e47282.

Chapter 7

1. Quote from p. 4 of Loren Cordain, *The Paleo Diet* (Boston, MA: Houghton Mifflin Harcourt, 2011).

2. Loren Cordain, *The Paleo Diet*.

3. John D. Speth, "Early Hominid Hunting and Scavenging: The Role of Meat as an Energy Source," *Journal of Human Evolution* 18 (1989) 329–343; Manual Domínguez-Rodrigo and Travis Rayne Pickering, "Early Hominid Hunting and Scavenging: A Zooarcheological Review," *Evolutionary Anthropology* 12 (2003) 275–282; Henry T. Bunn, "Meat Made Us Human," in *Evolution of the Human Diet*, ed. Peter S. Ungar (Oxford, UK: Oxford University Press, 2007), 191–211; Sujata Gupta, "Clever Eating," *Nature* 531 (March 2016): S12–13.

4. Peter S. Ungar, *Evolution of the Human Diet* (New York: Oxford University Press, 2007). Also see his popular science book, *Evolution's Bite: A Story of Teeth, Diet, and Human Origins* (Princeton, NJ: Princeton University Press, 2017).

5. Peter S. Ungar, "Dental Topography and Human Evolution with Comments on the Diets of *Australopithecus africanus* and *Paranthropus robustus*," in *Dental Perspectives on Human Evolution: State of the Art Research in Dental Paleoanthropology*, eds. S. Bailey and J.-J. Hublin (Dordrecht: Springer, 2007), 321–343.

6. Among the 15 species of fossil hominins for which molar enamel thickness has been described, only two are considered to have thin enamel relative to living humans: *Ardipithecus ramidus* and *Homo neanderthalensis*. Detailed further in Table 1, p. 396 of Smith et al., "Variation in Enamel Thickness Within the Genus *Homo*,"

Journal of Human Evolution 62 (2012): 395–411. This source also contains more information on Jolly and Kay's studies, including references.

7. Peter Andrews and Lawrence Martin, "Hominoid Dietary Evolution," *Philosophical Transactions: Biological Sciences*, 334, no. 1270 (November, 1991): 199–209.

8. Akiko Kato et al., "Intra- and Interspecific Variation in Macaque Molar Enamel Thickness," *American Journal of Physical Anthropology* 155 (2014): 447–459.

9. See, for example, Albert A. Dahlberg, "Dental Evolution and Culture," *Human Biology* 35, no. 3 (September, 1963): 237–249; James E. Anderson, "Skeletons of Tehuacán," *Science* 148, no. 3669 (April 1965): 496–497; Stephen Molnar, "Human Tooth Wear, Tooth Function and Cultural Variability," *American Journal of Physical Anthropology* 34 (1971): 175–190; B. Holly Smith, "Patterns of Molar Wear in Hunter-Gatherers and Agriculturalists," *American Journal of Physical Anthropology* 63 (1984): 39–56.

10. Examples include Mark F. Teaford, "Primate Dental Functional Morphology Revisited," in *Development, Function and Evolution of Teeth*, eds. Mark F. Teaford, Moya Meredith Smith, and Mark W. J. Ferguson (Cambridge, UK: Cambridge University Press, 2007), 290–304; Peter S. Ungar, "Mammalian Dental Function and Wear: A Review," *Biosurface and Biotribology* 1 (2015): 25–41; Mark F. Teaford and Alan Walker, "Dental Microwear in Adult and Still-Born Guinea Pigs (*Cavia porcellus*)," *Archives of Oral Biology* 28, no. 11 (1983): 1077–1081.

11. Experiments have shown that tiny features can appear after a single chewing cycle, particularly after eating hard foods, and tooth surfaces may be nearly completely reworked or replaced in a matter of days or weeks. M. F. Teaford and C. A. Tylenda, "A New Approach to the Study of Tooth Wear," *Journal of Dental Research* 70, no. 3 (1991): 204–207; Mark F. Teaford and Ordean J. Oyen, "In Vivo and in Vitro Turnover in Dental Microwear," *American Journal of Physical Anthropology* 80 (1989): 447–460; Mark F. Teaford and Kenneth E. Glander, "Dental Microwear in Live, Wild-Trapped *Alouatta palliata* from Costa Rica," *American Journal of Physical Anthropology* 85 (1991): 313–319.

12. Alan Walker, Hendrick N. Hoeck, and Linda Perez, "Microwear of Mammalian Teeth as an Indicator of Diet," *Science* 201, no. 4359 (September 1978): 908–910.

13. Mark F. Teaford and Jacqueline A. Runestad, "Dental Microwear and Diet in Venezuelan Primates," *American Journal of Physical Anthropology* 88 (1992): 347–364; Semprebon et al., "Can Low-Magnification Stereomicroscopy Reveal Diet?" *Journal of Human Evolution* 47 (2004): 115–144.

14. The discussion below is based on information from Scott et al., "Dental Microwear Texture Analysis Shows Within-Species Diet Variability in Fossil Hominins," *Nature* 436 (August 2005): 693–695; Peter S. Ungar and M. Sponheimer, "The Diets of Early Hominins," *Science* 334 (October 2011): 190–193; Peter S.

Ungar, Fredrick E. Grine, and Mark F. Teaford, "Dental Microwear and Diet of the Plio-Pleistocene Hominin *Paranthropus boisei*," *PLoS ONE* 3, no. 4 (2008): e2044; Peter S. Ungar, Fredrick E. Grine, Mark F. Teaford, and Sireen El Zaatari, "Dental Microwear and Diets of African Early *Homo*," *Journal of Human Evolution* 50 (2006): 78–95.

15. David Strait et al., "Viewpoints: Diet and Dietary Adaptations in Early Hominins: The Hard Food Perspective," *American Journal of Physical Anthropology* 151 (2013): 339–355.

16. Summarized in Steven E. Churchill, *Thin on the Ground: Neanderthal Biology, Archeology, and Ecology* (Ames, Iowa: John Wiley and Sons, 2014), 179–218.

17. Sireen El-Zaatari, "Occlusal Microwear Texture Analysis and the Diets of Historical/ Prehistoric Hunter-Gatherers," *International Journal of Osteoarcheology* 20 (2010): 67–87. However, see the lack of microwear formed by meat covered with abrasive grit in Li-Cheng Hua, Elizabeth T. Brandt, Jean-Francois Meullenet, Zhong-Rong Zhou, and Peter S. Ungar, "Technical Note: An In Vitro Study of Dental Microwear Formation Using the BITE Master II Chewing Machine," *American Journal of Physical Anthropology* 158 (2015): 769-775.

18. Mark F. Teaford and James D. Lytle, "Brief Communication: Diet-Induced Changes in Rates of Human Tooth Microwear: A Case Study Involving Stone-Ground Maize," *American Journal of Physical Anthropology* 100 (1996): 143–147.

19. M. A. Smith, "The Antiquity of Seedgrinding in Arid Australia," *Archaeology in Oceania* 21, no. 1 (1986): 29–39.

20. Herbert H. Covert and Richard F. Kay, "Dental Microwear and Diet: Implications for Determining the Feeding Behaviors of Extinct Primates, With a Comment on the Dietary Pattern of *Sivapithecus*," *American Journal of Physical Anthropology* 55 (1981): 331–336; Peter S. Ungar, Mark F. Teaford, Kenneth E. Glander, and Robert F. Pastor, "Dust Accumulation in the Canopy: A Potential Cause of Dental Microwear in Primates," *American Journal of Physical Anthropology* 97 (1995): 93–99.

21. Research on rabbits complicates things further, as adding abrasive particles to a uniform diet reduced microwear variation when compared to diets with few abrasives: see Ellen Schulz et al., "Dietary Abrasiveness Is Associated with Variability of Microwear and Dental Surface Texture in Rabbits," *PLoS ONE* 8, no. 2 (February 2013): e56167.

22. Peter W. Lucas, "Mechanisms and Causes of Wear in Tooth Enamel: Implications for Hominin Diets," *Journal of the Royal Society Interface* 10 (2013): 20120923; Li-Cheng Hua, Elizabeth T. Brandt, Jean-Francois Meullenet, Zhong-Rong Zhou, and Peter S. Ungar, "Technical Note: An In Vitro Study of Dental Microwear Formation Using the BITE Master II Chewing Machine"; Pia Nystrom, Jane E. Phillips-Conroy, and Clifford J. Jolly, "Dental Microwear in Anubis and Hybrid Baboons (*Papio hamadryas,*

Sensu Lato) Living in Awash National Park, Ethiopia," *American Journal of Physical Anthropology* 125 (2004): 279–291; Paul J. Constantino et al., "Tooth Chipping Can Reveal the Diet and Bite Forces of Fossil Hominins," *Biology Letters* 6 (2010): 826–829.

23. Fluoride is an exception, as it can increase in the outer enamel due to fluorination of water or clinical application. The mechanism of influx is not well understood.

24. Louise T. Humphrey, "Weaning Behaviour in Human Evolution," *Seminars in Cell and Developmental Biology* 21 (2010): 453–461; Christine Austin et al., "Barium Distributions in Teeth Reveal Early-Life Dietary Transitions in Primates," *Nature* 498 (2013): 216–219.

25. Tanya M. Smith, Christine Austin, Katie Hinde, Erin R. Vogel, and Manish Arora, "Cyclical Nursing Patterns in Wild Orangutans," *Science Advances* 3 (2017): e1601517.

26. Discussed further in D. W. Sellen, "Evolution of Human Lactation and Complementary Feeding: Implications for Understanding Contemporary Cross-cultural Variation," in *Breast-Feeding: Early Influences on Later Health*, eds. Gail Goldberg et al., (Netherlands: Springer, 2009), 253-82.

27. Christine Austin et al., "Barium Distributions in Teeth Reveal Early-Life Dietary Transitions in Primates." In full disclosure, it is much easier to study weaning in the teeth of living humans and primates than in fossil hominins. In addition to the challenges of getting permission to cut rare fossil teeth for mapping, it's crucial that the samples haven't been heavily modified after burial. Elements from the soil and groundwater often replace and obscure the original minerals in skeletal remains. The Belgian Neanderthal was in good shape; it retained enough original organic material to yield protein from the enamel and dentine, permitting other kinds of analyses discussed in chapter 8. My colleagues and I were quite lucky, as hominin fossils are rarely this well preserved!

28. The discussion below is based on information from Tim D. White et al., "Macrovertebrate Paleontology and the Pliocene Habitat of *Ardipithecus ramidus*," *Science* 326 (October 2009): 87–93; Matt Sponheimer et al., "Hominins, Sedges, and Termites: New Carbon Isotope Data from the Sterkfontein Valley and Kruger National Park," *Journal of Human Evolution* 48 (2005): 301–312; Thure E. Cerling et al., "Diet of *Paranthropus boisei* in the Early Pleistocene of East Africa," *Proceedings of the National Academy of Sciences USA* 108, no. 23 (June 2011): 9337–9341; Peter S. Ungar and Matt Sponheimer, "The Diets of Early Hominins," *Science* 334 (2011): 190–193; Henry et al., "The Diet of *Australopithecus sediba*," *Nature* 487 (July 2012): 90–93.

29. The unique and complex microwear signature of *Paranthropus robustus* is most similar to hard-object–eating primates, yet its ^{13}C values suggest a mixed C3-C4 diet similar to other hominins that did not appear to consume hard objects. In contrast, the simple striated pattern of *Paranthropus boisei* microwear implies that it ate

neither hard nor tough foods, while the ^{13}C signature is unlike that of any other hominin. Their extreme C4 signal is most similar to grass-eating warthogs, hippos, and zebras—pointing to a unique niche as a grazing hominin.

30. This debate is thoroughly discussed in the following sources: Frederick E. Grine et al., "Craniofacial Biomechanics and Functional and Dietary Inferences in Hominin Paleontology," *Journal of Human Evolution* 58 (2010): 293–308; David Strait et al., "Viewpoints: Diet and Dietary Adaptations in Early Hominins: The Hard Food Perspective"; Peter S. Ungar, Jessica R. Scott, Christine M. Steininger, "Dental Microwear Differences Between Eastern and Southern African Fossil Bovids and Hominins," *South African Journal of Science* 112, no. 3/4, art. #2015–0393 (2016); Justin A. Ledogar et al., "Mechanical Evidence That *Australopithecus sediba* Was Limited in Its Ability to Eat Hard Foods," *Nature Communications* 7 (2016): 10596.

31. There are subtle ^{13}C differences between plants in closed, forested environments and those in more open or dry environments, but they do not appear to distinguish grazers and browsers in France: see Michaela Ecker et al., "Middle Pleistocene Ecology and Neanderthal Subsistence: Insights from Stable Isotope Analysis in Payre (Ardèche, Southeastern France)," *Journal of Human Evolution* 65 (2013): 363–373.

32. Reviewed in Steven E. Churchill, *Thin on the Ground*. Molecular biologists have been successful in obtaining ancient DNA, enamel proteins, and enamel carbon ^{13}C values from older material.

33. Michael P. Richards and Erik Trinkaus, "Isotopic Evidence for the Diets of European Neanderthals and Early Modern Humans," *Proceedings of the National Academy of Sciences USA* 106, no. 38 (September 2009): 16034–16039.

34. Luca Fiorenza et al., "To Meat or Not to Meat? New Perspectives on Neanderthal Ecology," *Yearbook of Physical Anthropology* 156, suppl. S59 (2015): 43–71.

35. Yuichi I. Naito et al., "Ecological Niche of Neanderthals from Spy Cave Revealed by Nitrogen Isotopes of Individual Amino Acids in Collagen," *Journal of Human Evolution* 93 (April 2016): 82–90.

36. See, for example, Amanda G. Henry, "Recovering Dietary Information from Extant and Extinct Primates Using Plant Microremains," *International Journal of Primatology* 33 (2012): 702–715; Christina Warinner, Camilla Speller, Matthew J. Collins, and Cecil M. Lewis Jr., "Ancient Human Microbiomes," *Journal of Human Evolution* 79 (2015): 125–136.

37. Amanda G. Henry, Alison S. Brooks, and Dolores R. Piperno, "Microfossils in Calculus Demonstrate Consumption of Plants and Cooked Foods in Neanderthal Diets (Shanidar III, Iraq; Spy I and II, Belgium)," *Proceedings of the National Academy of Sciences USA* 108, no. 2 (January 2011): 486–491.

38. Karen Hardy et al., "Neanderthal Medics? Evidence for Food, Cooking, and Medicinal Plants Entrapped in Dental Calculus," *Naturwissenschaften* 99 (2012): 617–626.

39. Richard Wrangham, "The Cooking Enigma," in *Evolution of the Human Diet*, ed. Peter S. Ungar (New York: Oxford University Press, 2007), 308-323.

40. Chelsea Leonard, Layne Vashro, James F. O'Connell, and Amanda G. Henry, "Plant Microremains in Dental Calculus as a Record of Plant Consumption: A Test with Twe Forager-Horticulturalists," *Journal of Archaeological Science: Reports* 2 (2015): 449–457; Robert C. Power, Domingo C. Salazar-García, Roman M. Wittig, Martin Freiberg, Amanda G. Henry, "Dental Calculus Evidence of Taï Forest Chimpanzee Plant Consumption and Life History," *Scientific Reports* 5 (2015):15161.

41. Christina Warinner et al., "Direct Evidence of Milk Consumption from Ancient Human Dental Calculus," *Scientific Reports* 4 (2014): 7104.

42. Amanda G. Henry et al., "The Diet of *Australopithecus sediba*"; Justin A. Ledogar et al., "Mechanical Evidence That *Australopithecus sediba* Was Limited in Its Ability to Eat Hard Foods."

Chapter 8

1. Quoted from p. 172 of Albert A. Dalhberg, "Analysis of the American Indian Dentition," in *Dental Anthropology*, ed. Don R. Brothwell (New York: Pergamon Press, 1963).

2. Comment by C. Loring Brace on p. 397 in John A. Wallace, "Did La Ferrassie I Use His Teeth as a Tool?" *Current Anthropology* 16, no. 3 (September 1975): 393–401.

3. Examples are drawn from the following studies: Stephen Molnar, "Tooth Wear and Culture: A Survey of Tooth Functions Among Some Prehistoric Populations," *Current Anthropology* 13, no. 5 (December 1972): 511–526. Peter D. Shultz, "Task Activity and Anterior Tooth Grooving in Prehistoric Californian Indians," *American Journal of Physical Anthropology* 46 (1977): 87–92; John A. Wallace, "Did La Ferrassie I Use His Teeth as a Tool?"; Clark Spencer Larsen, "Dental Modifications and Tool Use in the Western Great Basin," *American Journal of Physical Anthropology* 67 (1985): 393–402; George R. Milner and Clark Spencer Larsen, "Teeth as Artifacts of Human Behavior: Intentional Mutilation and Accidental Modification," in *Advances in Dental Anthropology*, eds. Mark A. Kelley and Clarke S. Larsen (New York: Wiley-Liss, 1991), 357–378; Richard A. Gould, "Chipping Stones in the Outback," *Natural History* 77 (1968): 42–49; Inger Lous, "Om Mastikationsapparet Anvendt som Redskab [The Masticatory System as a Tool]," *Tandlaegebladet* 74, no. 1 (1970): 1–10; Kurt W. Alt and Sandra L. Pichler, "Artificial Modifications of Human Teeth," in *Dental Anthropology: Fundamentals, Limits, and Prospects*, eds. Kurt W. Alt, Friedrich W. Rösing, and Maria Teschler-Nicola (Vienna: Springer-Verlag, 1998), 387–415.

4. Bow drills are created by wrapping a taut bow string around a stick with a sharpened end, which is then rotated rapidly like a drill bit by moving the bow back and forth like a violin bow. This causes the sharpened end to spin, provided the free end is held in a socket, such as between slightly agape teeth. This technology was likely used to drill holes in teeth thousands of years ago, and is discussed in chapter 9.

5. John A. Wallace, "Did La Ferrassie I Use His Teeth as a Tool?"; Pierre-François Puech, "Tooth Wear in La Ferrassie Man," *Current Anthropology* 22, no. 4 (August 1981): 424–430; Peter S. Ungar, Karen J. Fennell, Kathleen Gordon, and Erik Trinkaus, "Neandertal Incisor Beveling," *Journal of Human Evolution* 32 (1997): 407–421.

6. Comment by C. Loring Brace on p. 396 of John A. Wallace, "Did La Ferrassie I Use His Teeth as a Tool?"

7. T. D. Stewart, "The Neanderthal Skeletal Remains from Shanidar Cave, Iraq: A Summary of Findings to Date," *Proceedings of the American Philosophical Society* 121, no. 2 (April 1977): 121–165.

8. José María Bermúdez de Castro, Timothy G. Bromage, and Yolanda Fernández Jalvo, "Buccal Striations on Fossil Human Anterior Teeth: Evidence of Handedness in the Middle and Early Upper Pleistocene," *Journal of Human Evolution* 17 (1988): 403–412; Marina Lozano, José M. Bermúdez de Castro, Eudald Carbonell, and Juan Luis Arsuaga, "Non-Masticatory Uses of Anterior Teeth of Sima de los Huesos Individuals (Sierra de Atapuerca, Spain)," *Journal of Human Evolution* 55 (2008): 713–728; Natalie T. Uomini, "Handedness in Neanderthals," in *Neanderthal Lifeways, Subsistence and Technology: One Hundred Fifty Years of Neanderthal Study*, eds. N. J. Conard and J. Richter (Dordrecht: Springer, 2011), 139–154.

9. Robert J. Hinton, "Form and Patterning of Anterior Tooth Wear Among Aboriginal Human Groups," *American Journal of Physical Anthropology* 54 (1981): 555–564; Peter S. Ungar, Karen J. Fennell, Kathleen Gordon, and Erik Trinkaus, "Neandertal Incisor Beveling."

10. Milford H. Wolpoff, "The Krapina Dental Remains," *American Journal of Physical Anthropology* 50 (1979): 67–114; Peter S. Ungar, Karen J. Fennell, Kathleen Gordon, and Erik Trinkaus, "Neandertal Incisor Beveling"; Adeline Le Cabec, Philipp Gunz, Kornelius Kupczik, José Braga, and Jean-Jacques Hublin, "Anterior Tooth Root Morphology and Size in Neanderthals: Taxonomic and Functional Implications," *Journal of Human Evolution* 64 (2013): 169–193.

11. Tanya M. Smith et al., "Variation in Enamel Thickness Within the Genus *Homo*," *Journal of Human Evolution* 62 (2012): 395–411; Tanya M. Smith, Kornelius Kupczik, Zarin Machanda, Matthew M. Skinner, and John P. Zermeno, "Enamel Thickness in Bornean and Sumatran Orangutan Dentitions," *American Journal of Physical Anthropology* 147 (2012): 417–426. This research has led me to wonder whether natural selection may have led to thickened enamel in anterior teeth that undergo routine compression

or high forces during biting. We don't yet know whether recent hunter-gatherers from the Pacific Northwest or Australia have thicker enamel or cementum on their front teeth than populations who don't engage in nondietary uses of teeth, but I'm hopeful that someone will tackle this question with nondestructive X-ray imaging.

12. Anna F. Clement, Simon W. Hillson, and Leslie C. Aiello, "Tooth Wear, Neanderthal Facial Morphology and the Anterior Dental Loading Hypothesis," *Journal of Human Evolution* 62 (2012): 367–376.

13. Reviewed in Steven R. Leigh, Joanna M. Setchell, Marie Charpentier, Leslie A. Knapp, and E. Jean Wickings, "Canine Tooth Size and Fitness in Male Mandrills (*Mandrillus sphinx*)," *Journal of Human Evolution* 55 (2008): 75–85.

14. Quote from p. 398 of Charles R. Darwin, *The Descent of Man, and Selection in Relation to Sex*, vol. 2 (London: John Murray, 1871).

15. Quote from ibid., 155.

16. Steven R. Leigh, Joanna M. Setchell, Marie Charpentier, Leslie A. Knapp, and E. Jean Wickings, "Canine Tooth Size and Fitness in Male Mandrills (*Mandrillus sphinx*)."

17. J. Michael Plavcan, "Sexual Size Dimorphism, Canine Dimorphism, and Male-Male Competition in Primates: Where Do Humans Fit In?" *Human Nature* 23 (2012): 45–67.

18. Canine dimorphism is moderate in common chimpanzees, who live in multi-male, multi-female groups. A band of related males form same-sex coalitions to defend territories inhabited by females through vocalizations, displays, and, in rare instances, physical violence. In this instance, strength in numbers appears to trump the need for an exaggerated "weapon for sexual strife"; see J. Michael Plavcan, Carel P. van Schaik, and Peter M. Kappeler, "Competition, Coalitions and Canine Size in Primates," *Journal of Human Evolution* 28 (1995): 245–276.

19. See measurements in Julius A. Keiser, *Human Adult Odontometrics* (Cambridge University Press, Cambridge, UK, 1990); J. Michael Plavcan, "Sexual Size Dimorphism, Canine Dimorphism, and Male-Male Competition in Primates"; Gen Suwa et al., "Paleobiological Implications of the *Ardipithecus ramidus* Dentition," *Science* 326 (October 2009): 94–99.

20. Alan J. Almquest, "Sexual Differences in the Anterior Dentition in African Primates," *American Journal of Physical Anthropology* 40 (1974): 359–368; Jay Kelley, "Sexual Dimorphism in Canine Shape Among Extant Great Apes," *American Journal of Physical Anthropology* 96 (1995): 365–389.

21. Gen Suwa et al., "Paleobiological Implications of the *Ardipithecus ramidus* Dentition."

22. C. Owen Lovejoy, "The Origin of Man," *Science* 211, no. 4480 (January 1981): 341–350.

23. M. D. Leakey and R. L. Hay, "Pliocene Footprints in the Laetolil Beds at Laetoli, Northern Tanzania," *Nature* 278 (March 1979): 317–323.

24. J. Michael Plavcan, "Sexual Size Dimorphism, Canine Dimorphism, and Male-Male Competition in Primates."

25. Jay Kelley, "Sexual Dimorphism in Canine Shape Among Extant Great Apes"; Brian Hare, Victoria Wobber, and Richard Wrangham, "The Self-Domestication Hypothesis: Evolution of Bonobo Psychology Is Due to Selection Against Aggression," *Animal Behavior* 83, no. 3 (March 2012): 573–585.

26. The story becomes more complicated when we consider sexual dimorphism in body size, which some scientists believe was greater in early hominins and australopithecines than dimorphism in canine size. Our current understanding of hominin body mass dimorphism is pretty nebulous, since there are several approaches for estimating body size from fossil remains, which often yield conflicting estimates. See discussion in J. Michael Plavcan, "Sexual Size Dimorphism, Canine Dimorphism, and Male-Male Competition in Primates."

27. Quote from p. 144 of Charles R. Darwin, *The Descent of Man, and Selection in Relation to Sex,* vol. 1 (London: John Murray, 1871).

28. Ralph L. Holloway, "Tools and Teeth: Some Speculations Regarding Canine Reduction," *American Anthropologist* 69 (1967): 63–67.

29. Robert L. Cieri, Steven E. Churchill, Robert G. Franciscus, Jingzhi Tan, and Brian Hare, "Craniofacial Feminization, Social Tolerance, and the Origins of Behavioral Modernity," *Current Anthropology* 55, no. 4 (August 2014): 419–443; Brian Hare, Victoria Wobber, and Richard Wrangham, "The Self-Domestication Hypothesis."

30. See table 12 in Jay Kelley, "Sexual Dimorphism in Canine Shape Among Extant Great Apes."

31. See, for example, M. R. Zingeser and C. H. Phoenix, "Metric Characteristics of the Canine Dental Complex in Prenatally Androgenized Female Rhesus Monkeys (*Macaca mulatta*)," *American Journal of Physical Anthropology* 49 (1978): 187–192; Tuomo Heikkinen, Virpi Harila, Juha S. Tapanainen, and Lassi Alvesalo, "Masculinization of the Eruption Pattern of Permanent Mandibular Canines in Opposite Sex Twin Girls," *American Journal of Physical Anthropology* 151 (2013): 566–572.

32. Wu Liu et al., "The Earliest Unequivocally Modern Humans in Southern China," *Nature* 526 (October 2015): 696–699; Kira E. Westaway et al., "An Early Modern Human Presence in Sumatra 73,000–63,000 Years Ago," *Nature* 548 (2017): 322–325. Paleoanthropologists have a special place in their hearts for priority—any time a fossil appears to be "the oldest" or "the first" it receives extra care and attention, which is reinforced by the popular science press and editorial preferences of elite scholarly journals. To be honest, a new fossil or artifact turns up every few years

to demonstrate that something happened earlier than we thought—usurping the significance of discoveries that had been sensationalized just a few years before.

33. See examples and discussion in G. Richard Scott and Christy G. Turner II, *The Anthropology of Modern Human Teeth: Dental Morphology and its Variation in Recent Human Populations* (Cambridge, UK: Cambridge University Press, 1997); Tsunehiko Hanihara, "Morphological Variation of Major Human Populations Based on Non-metric Dental Traits," *American Journal of Physical Anthropology* 136 (2008): 169–182; Christopher M. Stojanowski, Kent M. Johnson, and William N. Duncan, "Sinodonty and Beyond: Hemispheric, Regional, and Intracemetry Approaches to Studying Dental Morphological Variation in the New World," in *Anthropological Perspectives on Tooth Morphology: Genetics, Evolution, Variation*, eds. G. Richard Scott and Joel D. Irish (Cambridge, UK: Cambridge University Press, 2013), 408–452.

34. Michael Richards et al., "Strontium Isotope Evidence of Neanderthal Mobility at the Site of Lakonis, Greece Using Laser-Ablation PIMMS," *Journal of Archaeological Science* 35 (2008): 1251–1256.

35. Malte Willmes et al., "The IRHUM (Isotopic Reconstruction of Human Migration) Database—Bioavailable Strontium Isotope Ratios for Geochemical Fingerprinting in France," *Earth System Science Data* 6 (2014): 117–122.

36. Sandi R. Copeland et al., "Strontium Isotope Evidence for Landscape Use by Early Hominins," *Nature* 474 (2011): 76–78.

37. A subsequent study of South African hominins explored a different question about landscape use: Vincent Balter, José Braga, Philippe Télouk, and J. Francis Thackeray, "Evidence for Dietary Change but Not Landscape Use in South African Early Hominins," *Nature* 489 (2012): 558–560. The authors suggested that similarities in strontium isotope ratios in three groups *(Paranthropus robustus, Australopithecus africanus*, and early *Homo*) showed similar ranging behavior. While the average strontium values did not differ among these hominins, it would be interesting to know if there were differences between small- and large-toothed individuals, particularly as *Paranthropus robustus* showed a wide range of ratios. Moreover, although the authors state that home ranges were of similar size among the groups, these ratios are simply a reflection of the geology of where an individual spent a few years of their childhood. Adult male chimpanzees and orangutans often range more broadly than adult females with dependent offspring.

38. See, for example, Rainer Grün, "Direct Dating of Human Fossils," *Yearbook of Physical Anthropology* 49 (2006): 2–48.

39. Collagen contains stable and radioactive carbon isotopes that can be measured in specialized mass spectrometers. Radioactive ^{14}C undergoes a complex structural transition over time, becoming ^{14}N, a stable isotope of nitrogen. The amount of ^{14}C decreases by half every 5,730 years, which is known as the half-life of radiocarbon.

By measuring the amount of remaining ^{14}C and comparing it to the stable isotope ^{12}C, scientists can determine how long ago an individual died.

40. Reviewed in R. Grün and C. B. Stringer, "Electron Spin Resonance Dating and the Evolution of Modern Humans," *Archaeometry* 33, no. 2 (1991): 153–199; Rainer Grün, "Direct Dating of Human Fossils"; Mathieu Duval, "Electron Spin Resonance Dating of Fossil Tooth Enamel," in *Encyclopedia of Scientific Dating Methods*, eds. Jack W. Rink and Jeroen W. Thompson (Dordrecht, Netherlands: Springer, 2015), 239-246.

41. Nadin Rohland and Michael Hofreiter, "Ancient DNA Extraction from Bones and Teeth," *Nature Protocols* 2, no. 7 (2007): 1756–1762; C. J. Adler, W. Haak, D. Donlon, A. Cooper, and the Genographic Consortium, "Survival and Recovery of DNA from Ancient Teeth and Bones," *Journal of Archaeological Science* 38 (2011) 956–964; Denice Higgins and Jeremy J. Austin, "Teeth as a Source of DNA for Forensic Identification of Human Remains: A Review," *Science and Justice* 53 (2013) 433–441.

42. C. J. Adler, W. Haak, D. Donlon, A. Cooper, and the Genographic Consortium, "Survival and Recovery of DNA from Ancient Teeth and Bones."

43. Andrew T. Ozga et al., "Successful Enrichment and Recovery of Whole Mitochondrial Genomes from Ancient Human Dental Calculus," *American Journal of Physical Anthropology* 160 (2016): 220–228.

44. Ludovic Orlando et al., "Revisiting Neandertal Diversity with a 100,000 year old mtDNA Sequence," *Current Biology* 16, no. 11 (2006): R400–R402; T. M. Smith, M. Toussaint, D. J. Reid, A. J. Olejniczak, and J.-J. Hublin, "Rapid Dental Development in a Middle Paleolithic Belgian Neanderthal," *Proceedings of the National Academy of Sciences USA* 104 (2007): 20220–20225; Christina M. Nielsen-Marsh, et al., "Extraction and Sequencing of Human and Neanderthal Mature Enamel Proteins using MALDI-TOF/TOF MS," *Journal of Archaeological Science* 36 (2009): 1758–1763; Christine Austin et al., "Barium Distributions in Teeth Reveal Early Life Dietary Transitions in Primates," *Nature* 498 (2013): 216–219.

45. Carles Lalueza-Fox et al., "Genetic Evidence for Patrilocal Mating Behavior Among Neandertal Groups," *Proceedings of the National Academy of Sciences USA* 108 (2011): 250–253.

46. Susanna Sawyer et al., "Nuclear and Mitochondrial DNA Sequences from Two Denisovan Individuals," *Proceedings of the National Academy of Sciences USA* 112 (2015): 15696–15700; Viviane Slon et al., "A Fourth Denisovan Individual," *Science Advances* 3 (2017): e1700186.

47. For example, see the ESRF Paleontological Microtomographic Database featuring open access synchrotron X-ray scans of hominin fossils established by Paul Tafforeau: http://paleo.esrf.eu.

48. See, for example, Svante Pääbo, *Neanderthal Man: In Search of Lost Genomes* (New York: Basic Books, 2014); George H. Perry and Ludovic Orlando, "Ancient DNA and Human Evolution," *Journal of Human Evolution* 79 (2015): 1–3; Qiaomei Fu et al., "The Genetic History of Ice Age Europe," *Nature* 534 (June 2016): 200–205; David Reich, *Who We Are and How We Got Here: Ancient DNA and the New Science of the Human Past* (Oxford: Oxford University Press, 2018).

49. Rebecca Rogers Ackermann, Alex Mackay, and Michael L. Arnold, "The Hybrid Origin of 'Modern' Humans," *Evolutionary Biology* 43 (2016): 1–11.

Chapter 9

1. Javier Romero, "Dental Mutilation, Trephination, and Cranial Deformation," in *Handbook of Middle American Indians*, vol. 9., ed. T. Dale Stewart (Austin, TX: University of Texas Press, 1970), 50–67.

2. Gregorio Oxilia et al., "Earliest Evidence of Dental Caries Manipulation in the Late Upper Palaeolithic," *Scientific Reports* (2015): 12150.

3. It's not entirely clear why each tooth was drilled, since only four of the 11 teeth had evidence of decay. Might a few of these holes have been made for practice by the first dental students? Original reference: A. Coppa et al., "Early Neolithic Tradition of Dentistry," *Nature* 440 (2006): 755–756.

4. Federico Bernardini et al., "Beeswax as Dental Filling on a Neolithic Human Tooth," *PLOS ONE* 7, no. 9 (September 2012): e44904.

5. Apparently, dental extraction was uncommon or not employed at all in ancient Egypt, despite considerable evidence that dental disease was common during this time: see Roger J. Forshaw, "The Practice of Dentistry in Ancient Egypt," *British Dental Journal* 206, no. 9 (May 2009): 481–486.

6. This procedure is also referred to as evulsion or ablation, particularly in European literature. Overviews can be found in John C. Willman, Laura Shackelford, and Fabrice Demeter, "Incisor Ablation Among the Late Upper Paleolithic People of Tam Hang (Northern Laos): Social Identity, Mortuary Practice, and Oral Health," *American Journal of Physical Anthropology* 160 (2016): 519–528; Christopher M. Stojanowski, Kent M. Johnson, Kathleen S. Paul, and Charisse L. Carver, "Indicators of Idiosyncratic Behavior in the Dentition," in *A Companion to Dental Anthropology*, ed. Joel D. Irish and G. Richard Scott (West Sussex, UK: Wiley Blackwell), 377–395; Arthur C. Durband, Judith Littleton, and Keryn Walshe, "Patterns in Ritual Tooth Avulsion at Roonka," *American Journal of Physical Anthropology* 154 (2014): 479–485; George R. Milner and Clark Spencer Larsen, "Teeth as Artifacts of Human Behavior: Intentional Mutilation and Accidental Modification," in *Advances in Dental Anthropology*, eds. Mark A. Kelley and Clarke S. Larsen (New York: Wiley-Liss, 1991), 357–378;

Jim P. Mower, "Deliberate Ante-mortem Dental Modification and its Implications in Archaeology, Ethnography and Anthropology," *Papers from the Institute of Archaeology* 10 (1999): 37–53; Kurt W. Alt and Sandra L. Pichler, "Artificial Modifications of Human Teeth," in *Dental Anthropology: Fundamentals, Limits, and Prospects*, eds. Kurt W. Alt, Friedrich W. Rösing, and Maria Teschler-Nicola (Vienna: Springer-Verlag, 1998), 387–415.

7. Isabelle De Groote and Louise T. Humphrey, "Characterizing Evulsion in the Later Stone Age Maghreb: Age, Sex, and Effects on Mastication," *Quaternary International* 413 (2016): 50–61.

8. Steve Webb, *The Willandra Lakes Hominids* (Canberra: Research School of Pacific Studies, Australian National University, 1989), 66–67; James F. O'Connell and Jim Allen, "Pre-LGM Sahul (Pleistocene Australia-New Guinea) and the Archaeology of Early Modern Humans," in *Rethinking the Human Revolution*, eds. Paul Mellars, Katie Boyle, Ofer Bar-Yosef, and Chris Stringer (Cambridge, UK: Cambridge McDonald Institute Monographs, 2007), 395–410.

9. Archeological evidence currently supports the arrival of humans in Australia at least 65,000 years ago, but human skeletal remains from this time remain elusive. See Chris Clarkson et al., "Human occupation of northern Australia by 65,000 years ago," *Nature* 547 (2017): 306–310.

10. The convention "tooth mutilation" has been eschewed by recent scholars due to its potentially negative connotation in favor of "tooth modification." See Jim P. Mower, "Deliberate Ante-mortem Dental Modification and its Implications in Archaeology, Ethnography and Anthropology."

11. Overviews of this topic can be found in most of the references given in endnote 6.

12. B. C. Finucane, K. Manning, and M. Touré, "Prehistoric Dental Modification in West Africa—Early Evidence from Karkarichinkat Nord, Mali," *International Journal of Osteoarcheology* 18, no. 6 (2008): 632–640.

13. J. S. Handler, R. S. Corruccini, and R. J. Mutaw, "Tooth Mutilation in the Caribbean: Evidence from a Slave Burial Population in Barbados," *Journal of Human Evolution* 11 (1982): 297–313; Hannes Schroeder, Tamsin C. O'Connell, Jane A. Evans, Kristrina A. Shuler, and Robert E. M. Hedges, "Trans-Atlantic Slavery: Isotopic Evidence for Forced Migration to Barbados," *American Journal of Physical Anthropology* 139 (2009): 547–557.

14. Javier Romero, "Dental Mutilation, Trephination and Cranial Deformation"; Saúl Dufoo Olvera et al., "Decorados Dentales Prehispánicos [Pre-Hispanic Dental Decoration]," *Revista Odontológica Mexicana* 14, no. 2 (June 2010): 99–106; Christopher M. Stojanowski, Kent M. Johnson, Kathleen S. Paul, and Charisse L. Carver, "Indicators of Idiosyncratic Behavior in the Dentition."

15. Thomas J. Zumbroich and Analyn Salvador-Amores, "Gold Work, Filing and Blackened Teeth: Dental Modifications in Luzon," *The Cordillera Review* 2, no. 2 (2010): 3–42.

16. Overviews of this topic can be found in Jim P. Mower, "Deliberate Ante-mortem Dental Modification and its Implications in Archaeology, Ethnography and Anthropology"; A. Jones, "Dental Transfigurements in Borneo," *British Dental Journal* 191, no. 2 (July 2001): 98–102; Thomas J. Zumbroich, "'Teeth as Black as a Bumble Bee's Wings': The Ethnobotany of Teeth Blackening in Southeast Asia," *Ethnobotany Research & Applications* 7 (2009): 381–398; Christopher M. Stojanowski, Kent M. Johnson, Kathleen S. Paul, and Charisse L. Carver, "Indicators of Idiosyncratic Behavior in the Dentition"; Thomas J. Zumbroich, "'The *Missī*-Stained Finger-Tip of the Fair': A Cultural History of Teeth and Gum Blackening in South Asia," *eJournal of Indian Medicine* 8 (2015): 1–32; Rusyad Adi Suriyanto and Toetik Koesbardiati, "Dental Modifications: A Perspective of Indonesian Chronology and the Current Applications," *Dental Journal Majalah Kedokteran Gigi* 43, no. 2 (June 2010): 81-90.

17. Quote from p. 3 of Thomas J. Zumbroich, "'Teeth as Black as a Bumble Bee's Wings': The Ethnobotany of Teeth Blackening in Southeast Asia."

18. Records of dental practices in ancient Egypt suggest that numerous natural remedies were applied to attenuate pain or arrest disease; see F. Filce Leek, "The Practice of Dentistry in Ancient Egypt," *The Journal of Egyptian Archaeology* 53 (Dec., 1967): 51–58; R. J. Forshaw, "The Practice of Dentistry in Ancient Egypt." Intriguing historical accounts of dentistry can also be found in Sydney Garfield, *Teeth Teeth Teeth* (New York: Simon and Schuster, 1969) as well as through numerous online exhibits, such as the University of the Pacific's Virtual Dental Museum (http://www.virtualdentalmuseum.org). The American Academy of the History of Dentistry maintains a listing of numerous museums, libraries, archives, and historical societies.

19. Jeanne-Marie Granger and François Lévêque, "Castelperronian and Aurignacian Ornaments: A Comparative Study of Three Unexamined Series of Perforated Teeth," *Comptes Rendus de l'Académie des Sciences—Series IIA—Earth and Planetary Science* 325 (1997): 537–543; Randall White, "Systems of Personal Ornamentation in the Early Upper Palaeolithic: Methodological Challenges and New Observations," in *Rethinking the Human Revolution*, eds. Paul Mellars, Katie Boyle, Ofer Bar-Yosef, and Chris Stringer (Cambridge, UK: Cambridge McDonald Institute Monographs, 2007), 287–302.

20. Randall White, "Systems of Personal Ornamentation in the Early Upper Palaeolithic: Methodological Challenges and New Observations"; Matthew G. Leavesley, "A Shark-tooth Ornament from Pleistocene Sahul," *Antiquity* 81 (2007): 308–315; Lois Sherr Dubin. *The History of Beads: From 100,000 B.C. to the Present* (New York: Abrams, 2009).

21. Randall White, "Systems of Personal Ornamentation in the Early Upper Palaeolithic: Methodological Challenges and New Observations"; Fernando V. Ramirez

Rozzi et al., "Cutmarked Human Remains Bearing Neanderthal Features and Modern Human Remains Associated with the Aurignacian at Les Rois," *Journal of Anthropological Sciences* 87 (2009): 153–185.

22. Matthew G. Leavesley, "A Shark-tooth Ornament from Pleistocene Sahul," *Antiquity* 81 (2007): 308–315; Emanuela Cristiani, Ivana Zivaljevic, and Dusan Boric, "Residue Analysis and Ornament Suspension Techniques in Prehistory: Cyprinid Pharyngeal Teeth Beads from Late Mesolithic Burials at Vlasac (Serbia)," *Journal of Archaeological Science* 46 (2014): 292–310; Dominique Gambier, "Aurignacian Children and Mortuary Practice in Western Europe," *Anthropologie* 38, no. 1 (2000): 5–21.

23. Tom Higham et al., "The Timing and Spatiotemporal Patterning of Neanderthal Disappearance," *Nature* 512 (August 2014): 306–309.

24. Shara E. Bailey, Timothy D. Weaver, and Jean-Jacques Hublin, "Who Made the Aurignacian and Other Early Upper Paleolithic Industries?" *Journal of Human Evolution* 57 (2009): 11–26; Tom Higham et al., "The Timing and Spatiotemporal Patterning of Neanderthal Disappearance."

25. Shara E. Bailey and Jean-Jacques Hublin, "Dental Remains from the Grotte du Renne at Arcy-sur-Cure (Yonne)," *Journal of Human Evolution* 50 (2006): 485–508; François Caron, Francesco d'Errico, Pierre Del Moral, Frédéric Santos, and João Zilhão, "The Reality of Neandertal Symbolic Behavior at the Grotte du Renne, Arcy-sur-Cure, France," *PLOS ONE* 6, no. 6 (2011): e21545; Paul Mellars, "Neanderthal Symbolism and Ornament Manufacture: The Bursting of a Bubble?" *Proceedings of the National Academy of Sciences USA* 107, no. 47 (November, 2010): 20147–20148; François Caron, Francesco d'Errico, Pierre Del Moral, Frédéric Santos, and João Zilhão, "The Reality of Neandertal Symbolic Behavior at the Grotte du Renne, Arcy-sur-Cure, France"; Jean-Jacques Hublin et al., "Radiocarbon Dates from the Grotte du Renne and Saint-Césaire Support a Neandertal Origin for the Châtelperronian," *Proceedings of the National Academy of Sciences USA* 109, no. 46 (November 2012): 18743–18748.

26. Paul Mellars, "Neanderthal Symbolism and Ornament Manufacture: The Bursting of a Bubble?"; Jean-Jacques Hublin et al., "Radiocarbon Dates from the Grotte du Renne and Saint-Césaire Support a Neandertal Origin for the Châtelperronian."

27. Marie Soressi et al., "Neandertals Made the First Specialized Bone Tools in Europe," *Proceedings of the National Academy of Sciences USA* 110, no. 35 (August, 2013): 14186–14190; Dirk L. Hoffmann, Diego E. Angelucci, Valentín Villaverde, Josefina Zapata, and João Zilhão, "Symbolic Use of Marine Shells and Mineral Pigments by Iberian Neandertals 115,000 Years Ago," *Science Advances* 4, eaar5255 (February 2018), DOI: 10.1126/sciadv.aar5255

28. Qiaomei Fu et al., "An Early Modern Human from Romania with a Recent Neanderthal Ancestor," *Nature* 524 (August 2015): 216–219.

29. Paola Villa and Wil Roebroeks, "Neandertal Demise: An Archaeological Analysis of the Modern Human Superiority Complex" *PLOS ONE* 9, no. 4 (April 2014): e96424.

30. João Zilhão, "The Emergence of Ornaments and Art: An Archaeological Perspective on the Origins of "Behavioral Modernity," *Journal of Archaeological Research* 15, no. 1 (March 2007): 1–54.

31. Randall White, "Systems of Personal Ornamentation in the Early Upper Palaeolithic: Methodological Challenges and New Observations"; April Nowell, comment on Robert L. Cieri, Steven E. Churchill, Robert G. Franciscus, Jingzhi Tan, and Brian Hare, "Craniofacial Feminization, Social Tolerance, and the Origins of Behavioral Modernity," *Current Anthropology* 55, no. 4 (August 2014): 419–443.

32. N. W. G. Macintosh, K. N. Smith, and A. B. Bailey, "Lake Nitchie Skeleton—Unique Aboriginal Burial," *Archaeology & Physical Anthropology in Oceania* 5, no. 2 (July 1970): 85–101.

33. See https://news.nationalgeographic.com/news/2012/06/120627-worlds-oldest-purse-dog-teeth-science-handbag-friederich.

34. The teeth are L 894–1 RUP3 and LUP4, originally dated to 1.84 million years ago and attributed to *H. habilis*; and OH 60, attributed to *Homo erectus* and dated to 1.7–2.1 million years ago. Noel T. Boaz and F. Clark Howell, "A Gracile Hominid Cranium from Upper Member G of the Shungura Formation, Ethiopia," *American Journal of Physical Anthropology* 46 (1977): 93–107; Peter S. Ungar, Frederick E. Grine, Mark F. Teaford, and Alejandro Pérez- Pérez, "A Review of Interproximal Wear Grooves on Fossil Hominin Teeth with New Evidence from Olduvai Gorge," *Archives of Oral Biology* 46 (2001): 285–292.

35. Quote from p. 163 of Franz Weidenreich, "The Dentition of *Sinanthropus pekinensis*: A Comparative Odontography of the Hominids." In contrast to Weidenreich's position, some have gone so far as to infer that these marks proved that ancient hominins had the capacity to produce speech, and thus that human language originated 2.5 million years ago with early *Homo*: see William A. Agger, Timothy L. McAndrews, and John A. Hlaudy, "On Toothpicking in Early Hominids," *Current Anthropology* 45, no. 3 (June 2004): 403–404.

36. Leslea J. Hlusko, "The Oldest Hominid Habit? Experimental Evidence for Toothpicking with Grass Stalks," *Current Anthropology* 44, no. 5 (December 2003): 738–741.

37. William C. McGrew and Caroline E. G. Tutin, "Chimpanzee Dentistry," *The Journal of the American Dental Association* 85 (December 1972): 1198–1204; William C. McGrew and Caroline E. G. Tutin, "Chimpanzee Tool Use in Dental Grooming," *Nature* 241 (February 1973): 477–478; Jane Goodall, "Stanford Outdoor Primate Facility 'Gombe West,'" Kenneth M. Cuthbertson Papers (SC0582), Department of Special Collections and University Archives, Stanford University Libraries, box 40, folder 12 (ca. 1973).

38. Jane van Lawick-Goodall, "The Behaviour of Free-Living Chimpanzees in the Gombe Stream Reserve," *Animal Behavior Monographs* 1, part 3 (1968): 161–311; Sonya M. Kahlenberg and Richard W. Wrangham, "Sex Differences in Chimpanzees' Use of Sticks as Play Objects Resemble Those of Children," *Current Biology* 20, no. 24 (2010): R1067–1068.

39. Examples given in this passage derive from Jean-Baptiste Leca, Noëlle Gunst, and Michael A. Huffman, "The First Case of Dental Flossing by a Japanese Macaque (*Macaca fuscata*): Implications for the Determinants of Behavioral Innovation and the Constraints on Social Transmission," *Primates* 51 (2010): 13–22; Kunio Watanabe, Nontakorn Urasopon, and Suchinda Malaivijitnond, "Long-Tailed Macaques Use Human Hair as Dental Floss," *American Journal of Primatology* 69 (2007): 940–944; Nobuo Masataka, Hiroki Koda, Nontakorn Urasopon, and Kunio Watanabe, "Free-Ranging Macaque Mothers Exaggerate Tool-Using Behavior when Observed by Offspring," *PLOS ONE* 4, no. 3 (March 2009): e4768; Ellen J. Ingmanson, "Tool-Using Behavior in Wild *Pan paniscus*: Social and Ecological Considerations," in *Reaching Into Thought: The Minds of the Great Apes*, eds. Anne E. Russon, Kim A. Bard, and Sue Taylor Parker (Cambridge, UK: Cambridge University Press, 1996), 190–210; Anne E. Russon et al., "Innovation and Intelligence in Orangutans," in *Orangutans: Geographic Variation in Behavioral Ecology and Conservation*, eds. Serge A. Wich, S. Suci Utami Atmoko, Tatang Mitra Setia, and Carel P. van Schaik (New York: Oxford University Press, 2009), 279–298; Jane van Lawick-Goodall, "The Behaviour of Free-Living Chimpanzees in the Gombe Stream Reserve"; Gordon G. Gallup, Jr., "Chimpanzees: Self-Recognition," *Science* 167, No 3914 (Jan 1970): 86–87; *Daily Mail*, October 12, 2015, http://www.dailymail.co.uk/sciencetech/article-3269353/Dental-hygiene-monkey-business-Baboon-spotted-FLOSSING-teeth-using-bristles-broom.html.

40. B. Bonfigliolo, V. Mariotti, F. Facchini, M.G. Belcastro, and S. Condemi, "Masticatory and Non-masticatory Dental Modifications in the Epipalaeolithic Necropolis of Taforalt (Morocco)," *International Journal of Osteoarcheology* 14 (2004): 448–456; Ann Margvelashvili, Christoph P. E. Zollikofer, David Lordkipanidze, Timo Peltomäki, and Marcia S. Ponce de León, "Tooth Wear and Dentoalveolar Remodeling are Key Factors of Morphological Variation in the Dmanisi Mandibles," *Proceedings of the National Academy of Sciences USA* 110, no. 43 (October 2013): 17278–17283; Marina Lozano, Maria Eulàlia Subirà, José Aparicio, Carlos Lorenzo, and Gala Gómez-Merino, "Toothpicking and Periodontal Disease in a Neanderthal Specimen from Cova Foradà Site (Valencia, Spain)," *PLOS ONE* 8, no. 10 (October 2013): e76852.

Concluding Thoughts: The Future That Teeth Foretell

1. Sarang Sharma, Dhirendra Srivastava, Shibani Grover, and Vivek Sharma, "Biomaterials in Tooth Tissue Engineering: A Review," *Journal of Clinical and Diagnostic Research* 8, no. 1 (2014): 309–315; Ajaykumar Vishwakarma, Paul Sharpe, Songtao

Shi, and Murugan Ramalingam, eds., *Stem Cell Biology and Tissue Engineering in Dental Sciences* (London: Elsevier, 2015); Anne Baudry, Emel Uzunoglu, Benoit Schneider, Odile Kellermann, and Michel Goldberg, "From Pulpal Stem Cells to Tooth Repair: An Emerging Field for Dental Tissue Engineering," *Evidence-Based Endodontics* 1 (2016): 2.

2. The Stem Cell Australia research collective lists 18 ongoing stem cell therapy clinical trials that are actively recruiting patients as of June 2017; see http://www.stemcellsaustralia.edu.au/About-Stem-Cells/trials-in-australia.aspx.

3. For more information, see Andrew H. Jheon, Kerstin Seidel, Brian Biehs, and Ophir D. Klein, "From Molecules to Mastication: The Development and Evolution of Teeth," *WIREs Developmental Biology* (May 3, 2012), doi: 10.1002/wdev.63. A useful animation of this process can be seen here: https://www.youtube.com/watch?time_continue=1&v=ozyKNVabfos

4. For more information, see Paul T. Sharpe, "Dental Mesenchymal Stem Cells," *Development* 143 (2016): 2273–2280; Anne Baudry, Emel Uzunoglu, Benoit Schneider, Odile Kellermann, and Michel Goldberg, "From Pulpal Stem Cells to Tooth Repair: An Emerging Field for Dental Tissue Engineering."

5. Vitor C. M. Neves, Rebecca Babb, Dhivya Chandrasekaran, and Paul T. Sharpe, "Promotion of Natural Tooth Repair by Small Molecule GSK3 Antagonists," *Scientific Reports* 7 (2017): 39654.

6. Masako Miura, "SHED: Stem Cells from Human Exfoliated Deciduous Teeth," *Proceedings of the National Academy of Sciences USA* 100 (2003): 5807–5812. A discussion of the value of banking these teeth can be found here: http://www.cnn.com/2017/04/26/health/dental-stem-cell-banking/index.html.

7. A group of researchers led by Paul Sharpe combined stem cells with dental epithelium in mice, which begins the molecular cascade of tooth formation. They were able to do this in lab conditions outside the body, as well as in mouse kidneys, where the capsule around the organ provides a more natural environment. In three cases, they were able to grow recognizable mice teeth from the combination of bone marrow cells and dental epithelium. See Sonie A. C. Modino and Paul T. Sharpe, "Tissue Engineering of Teeth Using Adult Stem Cells," *Archives of Oral Biology* 50 (2005): 255–258. More recent studies have been rather cautionary about these results, as in Nelson Monteiro and Pamela C. Yelick, "Advances and Perspectives in Tooth Tissue Engineering," *Journal of Tissue Engineering and Regenerative Medicine* 11, no. 9 (2016): 2443, doi: 10.1002/term; Paul T. Sharpe, "Dental Mesenchymal Stem Cells."

8. Takashi Tsuji, "Bioengineering of Functional Teeth," in *Stem Cells in Craniofacial Development and Regeneration*, eds. George T.-J. Huang and Irma Thesleff (Hoboken, NJ: Wiley-Blackwell, 2013), 447–459.

9. Elizabeth E. Smith et al., "Developing a Biomimetic Tooth Bud Model," *Journal of Tissue Engineering and Regenerative Medicine* 10 (2017): 1–11; Atsuhiko Hikita,

Ung-il Chung, Kazuto Hoshi, and Tsuyoshi Takato, "Bone Regenerative Medicine in Oral and Maxillofacial Region Using a Three-Dimensional Printer," *Tissue Engineering: Part A* 23 (2017): 515–521. Scaffolds are constructed with natural and synthetic materials to create surfaces for cells to grow and interact, and they can be produced with special 3D printers. This additive manufacturing technology creates 3D objects from virtual computer models by adding solid layers upon one another in specialized printers. New applications seem to be developed constantly, including 3D-printed automotive parts, clothing, and even food! In one example, a combination of medical imaging and 3D bioprinting has helped researchers create customized bone implants for human patients. My Internet search for "3D-printed bone" brought up news coverage of several successful implantation surgeries over the past year. The truth is that it's much easier to bioengineer bones than teeth, since tooth production requires the interaction of several different embryonic cell types in a carefully timed sequence. And unlike bone, teeth don't remodel after their initial mineralization, so the placement and activation of secretory cells is a crucial aspect of their formation.

10. While studies of human twins have pointed to genetic underpinnings for certain visual impairments, the details of how these genes may have changed over time are only slowly coming into focus. The environmental determinants of poor vision are discussed in Daniel E. Lieberman, *The Story of the Human Body: Evolution, Health, and Disease* (New York, NY: Vintage Books, 2013)

11. Toby E. Hughes and Grant C. Townsend, "Twin and Family Studies of Human Dental Crown Morphology: Genetic, Epigenetic, and Environmental Determinants of the Modern Human Dentition," in *Anthropological Perspectives on Tooth Morphology*, eds. G. Richard Scott and Joel D. Irish (Cambridge, UK: Cambridge University Press, 2013), 31–68.

12. Sean G. Byars, Douglas Ewbank, Diddahally R. Govindaraju, and Stephen C. Stearns, "Natural Selection in a Contemporary Human Population," *Proceedings of the National Academy of Sciences USA* 107 (2010): 1787–1792.

13. Katherine M. Kirk, "Natural Selection and Quantitative Genetics of Life-History Traits in Western Women: A Twin Study," *Evolution* 55, no. 2 (2001): 423–435; Emmanuel Milot et al., "Evidence for Evolution in Response to Natural Selection in a Contemporary Human Population," *Proceedings of the National Academy of Sciences USA* 108, no. 41 (2011): 17040–17045.

14. Discussed further in Natalie R. Langley and Sandra Cridlin, "Changes in Clavicle Length and Maturation in Americans: 1840–1980," *Human Biology* 88, no. 1 (2016): 76–83.

15. Hans-Peter Kohler, Joseph L. Rodgers, and Kaare Christensen, "Is Fertility Behavior in Our Genes? Findings from a Danish Twin Study," *Population and Development Review* 25, no. 2 (1999): 253–288.

16. C. L. B. Lavelle, "Variation in the Secular Changes in the Teeth and Dental Arches," *Angle Orthodontist* 43 (1973): 412–421; Edward F. Harris, Rosario H. Potter, and Jiuxiang Lin, "Secular Trend in Tooth Size in Urban Chinese Assessed From Two-Generation Family Data," *American Journal of Physical Anthropology* 115 (2001): 312–318.

17. Natalie R. Langley and Sandra Cridlin, "Changes in Clavicle Length and Maturation in Americans: 1840–1980."

18. See, for example, Hugo F. V. Cardoso, Yann Heuzé, and Paula Júlio, "Secular Change in the Timing of Dental Root Maturation in Portuguese Boys and Girls," *American Journal of Human Biology* 22 (2010): 791–800; Anja Sasso et al., "Secular Trend in the Development of Permanent Teeth in a Population of Istria and the Littoral Region of Croatia," *Journal of Forensic Science* 58, no. 3 (2013): 673–677; J. Jayaraman, H. M. Wong, N. King, G. Roberts, "Secular Trends in the Maturation of Permanent Teeth in 5 to 6 Years Old Children," *American Journal of Human Biology* 25, no. 3 (2013): 329–334; Strahinja Vucic et al., "Secular Trend of Dental Development in Dutch Children," *American Journal of Physical Anthropology* 155 (2014): 91–98.

19. Some have argued that this acceleration may be partially due to socioeconomic factors, as Portuguese children from more affluent backgrounds show earlier third molar mineralization than their less-advantaged peers: see J. L. Carneiro, I. M. Caldas, A. Afonso, and H. F. V. Cardoso, "Examining the Socioeconomic Effects on Third Molar Maturation in a Portuguese Sample of Children, Adolescents and Young Adults," *International Journal of Legal Medicine* (2017) 131: 235–242. However, this study didn't control for jaw size, so it isn't clear whether this might have led to the differences in third molar calcification.

20. For example, see Christina Warinner, "Dental Calculus and the Evolution of the Human Oral Microbiome," *California Dental Association Journal* 44 (2016): 411–420; Andres Gomez et al., "Host Genetic Control of the Oral Microbiome in Health and Disease," *Cell Host & Microbe* 22 (2017): 269–278.

21. I. M. Porto et al., "Recovery and Identification of Mature Enamel Proteins in Ancient Teeth," *European Journal of Oral Sciences* 119 (2011): 83–87; G. A. Castiblanco et al., "Identification of Proteins from Human Permanent Erupted Enamel," *European Journal of Oral Sciences* 123 (2015): 390–395; Nicolas Andre Stewart et al., "The Identification of Peptides by NanoLC-MS/MS from Human Surface Tooth Enamel Following a Simple Acid Etch Extraction," *Royal Society of Chemistry Advances* 6 (2016): 61673–61679.

22. Reviewed in Manish Arora and Christine Austin, "Teeth as a Biomarker of Past Chemical Exposure," *Current Opinion Pediatrics* 25 (2013): 261–267; Guiqiang Liang et al., "Manganese Accumulation in Hair and Teeth as a Biomarker of Manganese Exposure and Neurotoxicity in Rats," *Environmental Science and Pollution Research* 23, no. 12 (2016): 12265–12271.

23. Louise Zibold Reiss, "Strontium-90 Absorption by Deciduous Teeth," *Science, New Series* 134, no. 3491 (Nov. 24, 1961): 1669–1673; Harold L. Rosenthal, John E. Gilster, and John T. Bird, "Strontium-90 Content of Deciduous Human Incisors," *Science, New Series* 140, no. 3563 (Apr. 12, 1963): 176–177; also see discussion on The Pauling Blog: https://paulingblog.wordpress.com/2011/06/01/the-baby-tooth-survey.

Index